# A BIRDWATCHERS' GUIDE TO CUBA, JAMAICA, HISPANIOLA, PUERTO RICO AND THE CAYMANS

**by Guy Kirwan, Arturo Kirkconnell and Mike Flieg**

Illustrations by Tony Disley

Prion Ltd.
Cley

# ABOUT THE AUTHORS

Guy Kirwan is a freelance ornithological editor and tour-leader who lives a dual life, with bases in Norwich, UK, and Rio de Janeiro, Brazil, a country where he has travelled particularly extensively. He is the editor of the *Bulletin of the British Ornithologists' Club*, one of the longest-running ornithological journals, and has edited a host of books on birds and wildlife. Guy has published widely on Turkish birds and is the main author of a recently published monograph on the country's avifauna, in addition to being a regular contributor to the technical literature on Neotropical birds.

Arturo Kirkconnell has for many years been the curator of birds at the Natural History Museum in Havana, Cuba, and he is the co-author of a *Field Guide to the Birds of Cuba* (published by Cornell University Press and Christopher Helm). A popular tour-leader for many years, numerous birdwatching visitors to the island have enjoyed his good humour and company. His research into Cuban birds, especially their ecology and taxonomy, has taken him not only throughout his native island but also to museums in the United States of America and England, and has led to numerous publications in the scientific literature. Arturo is married and has two children.

G. Michael Flieg has travelled extensively throughout the Caribbean, including to all of the islands in the Greater Antilles. He is the author of the recently published *A Photographic Guide to the Birds of the West Indies* (New Holland, 2000), and has organised numerous birdwatching trips throughout the world, including several to Cuba, one result of which was the meeting of the three authors of this book. Mike is a former Curator of Birds at both the St. Louis and Brookfield (Chicago) zoos. Elected to the Avicultural Hall of Fame, he has several first captive breeding records to his credit, and has published in avicultural, ornithological and veterinary journals. Founder of Ornifolks birding group, he resides in St. Louis, Missouri, with his wife of 50 years, Jacqueline. Grouse are his current passion.

# CONTENTS

# ACKNOWLEDGEMENTS

We are grateful to the following who have passed records, posted trip reports to various Internet birding sites or offered other information for inclusion in this book: Roger Ahlman, Chris Balchin, John Bell, Keith Betton, Chris Bradshaw, Sue Capewell, José Colón, Matt Denton, Kevin Easley, Göran Engberg, Daphne Gemmill, Phil Gibson, Derek Gruar, Henk Hendriks, Alex Kirschel, Miguel Landestoy, Magnus Liljefors, Cathy MacLaggan, Tim Marlow, John Martin, Blake Maybank, Jonathan Newman, Mark Oberle, Stefan Oscarsson, Ramona Oviedo, Lars Petersson, Bill & Rowena Quantrill, Malcolm Roxby, Dave & Nad Sargeant, Steve Smith, William Suárez, Mark Sutton, Lars Thornberg, Rick Waldrop, Susan Wallis, Rob Williams, John van der Woude and Barry Wright. We thank Richard Webb for preparing the mammal list.

The RARE Center for Tropical Conservation supplied some of the equipment and materials used to prepare the text, and the cooperation of the following authorities was also invaluable: BirdLife International, Comisión Nacional para la Protección de la Flora y la Fauna (Cuba), Ministerio de Agricultura, Empresa Municipal Agropecuaria Victoria de Girón de Cuba, Dirección de Silvicultura del Ministerio de Agricultura (Cuba), Instituto de Investigaciones Forestales (Cuba), Museo Nacional de Historia Natural y Instituto de Ecología y Sistemática de la Academia de Ciencias de Cuba, and Cubatur, Havanatour and Marazul tourism agencies. Graeme Gibson provided AK with opportunities to travel through Cuba, by organising specialised bird tours in the early years. Many local guides, drivers and tourist guides helped make these (and subsequent) tours a success. AK is grateful to all the colleagues and birders that have supported this project, which was initially conceived as a birding guide to Cuba alone, especially Orlando Garrido, Osmany González, Emilio Alfaro, George Wallace, Jim Wiley, Davis Finch, Allan Keith, Andy Mitchell, and Dana & Bob Fox. Esteban Gutiérrez greatly assisted in producing the maps.

Probably the person that stimulated Arturo's interest most was Rogelio García (El Pelao), whose skill and desire to find every single bird during his formative years as a birdwatcher has been of everlasting importance in his development as a tour leader. Arturo takes the greatest pleasure in dedicating the book to his memory. The work of the Birders' Exchange (a programme of the American Birding Association) in providing materials and equipment for bird conservation and research in Latin America and the Caribbean is an invaluable facility, from which Arturo has benefited on several occasions. He takes this opportunity to acknowledge the work of the programme's director, Betty Petersen, and all of those people and organisations that have and continue to donate equipment to Birders' Exchange. Finally, as always, Arturo expresses his immense gratitude to his wife Rosita for all her amazing support at home, which was of inestimable value in making this book possible.

GMK is grateful to a large number of individuals, especially Andy Mitchell who stimulated and nurtured his enthusiasm. Mike Flieg organised my first trip to Cuba, and subsequently he and Mark Elwonger furnished me with many opportunities to return to the region. Chris Bradshaw and Rob Williams were first-rate companions on that first, great, yet crazy trip to the Dominican Republic. Duncan Brooks joined a memorable but whirlwind trip to Cuba, the Dominican Republic and Puerto Rico. Many other friends have joined me in the field in the Greater Antilles, too many to mention by name, but I thank them all for sharing some great times. Hadoram Shirihai entirely financed my most recent return to the region, in spring 2009, and we are both most grateful to Ricardo Valentin for affording us special, short-notice access to the Puerto Rican Parrot reintroduction programme at Río Abajo State Forest, and for so generously sharing his time and knowledge with us. Storrs Olson and David Wege were more than kind in providing some literature at short notice prior to our publication deadline. During that period of my life that I have been working in Cuba, I have been more than fortunate to spend a significant proportion of my time in Brazil. Some great debts of gratitude must be paid to my 'Brazilian' friends, especially: Jeremy Minns and family, Fernando Pacheco, Eduardo Moreira Santos and Ana Lúcia de Almeida Faria, and, above all, Veronica Nogueira Gama, who even joined one trip to Cuba, as well as keeping the 'home fires' burning during the many periods when I was in the field.

Mike Flieg is grateful initially to the Caribbean birding pioneers, Ro Wauer and Craig Faanes, who graciously published their trip reports for those of us who followed. Allan Sander and Todd Mark accompanied him on many trips. Allan provided much information on Jamaica and other areas, as he pursued his photographic interests. Dave Klauber's trip reports have been extremely helpful. Joel Abramson gave me the opportunity to explore the Lesser Antilles, whilst Guy Kirwan invited me to explore eastern Cuba. My wife, Jackie, accompanied me initially, and what began as Caribbean vacations sometimes became birding nightmares for her. She handled them admirably.

# INTRODUCTION

In comparison to many other parts of the New World tropics, the Greater Antilles is not a particularly species-rich region. To date, the five archipelagos and six countries that are treated here as comprising the Greater Antilles can muster records of only just over 550 species, of which more than 5% are introductions (mainly to the US territory of Puerto Rico). Unsurprisingly, the largest island, Cuba, has the largest avifauna of the group, at about 360 species, while the tiny Caymans have played host to just over 220 species. However, when it comes to endemic birds, for which the region is justly famed and which provide the main impetus for birders to visit, Cuba's land area no longer proves to be a dominant factor. In common with the most recent review of its avifauna, we recognise 31 species as being endemic to Hispaniola (Dominican Republic and Haiti combined), of which only one, Grey-crowned Palm Tanager, is (probably) restricted to little-visited Haiti. Another species, Golden Swallow, may now be extirpated on Jamaica and therefore a *de facto* Hispaniolan endemic. Jamaica is now deemed to possess 30 endemic species (of which two might prove to be extinct), Cuba has 28 (of which one became extinct in the 19th century and another, Giant Kingbird, formerly occurred on two of the Bahamas), Puerto Rico has a respectable 18 and the Caymans none, although several endemic subspecies occur and Vitelline Warbler is most easily found on the islands. (In considering endemism, we disregard introductions to additional islands.)

With 105 single-island endemic species extant, the Greater Antilles offer far more than just the famous tropical sun, sea and sand that entice most visitors to their shores. Endemism in other groups is even more exceptional. For instance, on Cuba, where fewer than 10% of the birds are unique to the archipelago, 70% of its 54 mammals are endemic, 90% of both its reptiles and amphibians, 60% of its freshwater fish and the spiders and allies, while even 17% of its butterflies are restricted to the group. Furthermore, significant biodiversity remains uncatalogued: one very recent paper described seven new species of spiders in the same genus, all of them endemic to Cuba.

The islands are also of great importance for biodiversity conservation, especially birds. In 2000, when BirdLife International published their *Threatened birds of the world*, the Caymans held three species of conservation concern, Cuba 23, Dominican Republic 21, Jamaica 15 and Puerto Rico has ten. Furthermore, BirdLife recognises four Endemic Bird Areas within the region, of which two, Cuba and Hispaniola, were accorded the highest level of importance. At the time of writing, BirdLife's staff is in the final throes of publishing an Important Bird Area (IBA) inventory for the Caribbean region. Twenty-eight such areas have been identified on Cuba alone, while the Dominican Republic and Puerto Rico each boast 20 IBAs, Jamaica 15 and Haiti and the Caymans ten apiece.

Collectively, we have seen all of the extant endemics in the region. AK and GMF have visited the Caymans on several occasions, GMF and GMK have visited all of the other islands in the region multiple times, and between them AK and GMK have particularly extensive field experience on Cuba, where they have guided visiting birdwatchers on many occasions, as a largely pleasant diversion from their work in cataloguing, understanding more of, and most importantly conserving the Cuban avifauna.

We have assumed, above all else, that a guide of this nature must be designed around assisting the visitor, perhaps with less than two weeks at his/her disposal, to locate the endemic species for which the islands are justly famed. In each country's introduction we have made some suggestions as to time necessary for combining different islands on the same trip. However, we have also attempted to include interesting localities from most areas of each island, so that birders perhaps restricted by the constraints of a business or family visit might enjoy a rewarding locale reasonably close to their base. Of course, such an aim has not always been possible or practicable, but we have done our best to be as complete as possible within certain constraints and parameters.

Most birding visitors to Cuba visit only the western two-thirds of the island, and indeed our own field work in the eastern third of the country (the Oriente) has been comparatively limited, and undoubtedly other great areas for birding await discovery, particularly in this part of the country; those with time at their disposal are urged to contribute to our understanding of bird distributions in Cuba, or any of the other islands covered here by getting off the beaten track. We would be most interested to hear from users of this guide who do venture away from the standard circuit, and welcome their feedback (see below). Observations of globally threatened species should, of course, also be submitted to BirdLife International, Wellbrook Court, Girton Road, Cambridge CB3 0NA, UK, and, if possible, to the local BirdLife partner organisation.

In writing our introductory sections, we have been conscious of the need to restrict ourselves to the absolute essentials, those things that we consider a *birding* visitor *needs* to know. All of us are well aware that birders are not particularly accomplished readers of introductions, their desire to get to the most important parts of a book appears to overtake the urge to understand what is contained within its covers. In any case, as we point out, there are several very good general travel guides to the islands of the region and we have attempted to duplicate only those nuggets of information we consider particularly important or valuable to the birder. Thus, you will generally not find long lists of recommended hotels or their relative costs here, such information is readily available elsewhere, and is always subject to change. In the latter respect, we are also painfully aware that almost any bird book is out of date before it is even published. To this end, we welcome significant updates to the information presented here, which should be sent to the authors (via e-mail: GMKirwan@aol.com), for possible inclusion in any future editions of this guide. We are also considering the possibility of making particularly important updates available online via the first author's website.

# PRE-TOUR INFORMATION

*Visas and passports*

Red tape is largely very minimal in the Greater Antilles. US travellers are not even required to carry their passport to visit Puerto Rico, and US residents may travel in and out of Puerto Rico without going through customs or immigration. Other potential visitors are bound by the same rules as those travelling to the USA, making it important that visitors from countries, like the UK, bound by the visa waiver programme complete the online ETSA form prior to travel. There are no health restrictions, nor precautions. US Customs enforce Puerto Rico's entry and exit requirements. North American and EU visitors to other islands in the region require nothing more than a valid, current passport. Short-stay tourist visitors (intending to remain on any of the islands for less than one month) do not require visas, but a tourist card (valid for 90 days) must be purchased and completed on arrival in the Dominican Republic. No inoculations are required to enter any of the islands of the Greater Antilles from Europe, North America, Australia or New Zealand.

The only 'special' situation pertains to travel to Cuba. Most travellers to Cuba require only a valid passport, return ticket and 30-day tourist card in order to visit the country. (The situation for would-be US visitors is outlined in the separate section below.) Most of the specialist Cuba tour agencies (such as Havanatour) issue the tourist card as part of your air ticket price, and if you are travelling from another of the Greater Antilles, you should be able to purchase the card at your airline's check-in desk. However, at least in the UK, most regular travel agencies do not issue the tourist card with a ticket, which means you should obtain it from your nearest Cuban Embassy or Consulate. In the UK, this is at 167 High Holborn, London WC1 6PA (telephone: 0207 240288). The cost is £15 in the UK but can be cheaper elsewhere (e.g. currently US$15 when joining a Cuba-bound flight from Jamaica or Dominican Republic). Note that when completing the tourist card you need to designate a hotel in which you are going to stay during some of your time on the island. Customs officials in Cuba do not stamp your passport but the tourist card, whatever your nationality, unless you request that they do so. There is also an airport tax (currently US$25), payable on departure from Cuba.

*US travellers to Cuba*

For the purposes of normal tourism, the US government does not permit its citizens to travel to Cuba, as part of the ongoing embargo aimed at destabilising the Castro regime. Travel-related sanctions are administered by the Treasury Department's Office of Foreign Assets Control (OFAC). Currently, there is a total of 12 categories of activities for which travel may be sanctioned by the US authorities, under either a general or specific license; included within the former category is diplomacy, full-time journalism, certain types of professional meetings, educational research, and family visits. Applications under a specific license are considered on a case-by-case basis and must fall under one of the following: educational exchanges not involving academic study towards a degree programme, participation in a public performance, workshop or exhibition, support for the Cuban people through humanitarian efforts that will have no positive benefit to the Castro government, and the facilitation of certain exports. However, any activity that falls outside of those listed is considered 'tourism' and therefore subject to a strict ban on the granting of licenses and therefore the requisite visas. As made clear in the OFAC information concerning restrictions on travel to Cuba, and despite a common misconception to the contrary, certain speciality tours to Cuba where all costs are pre-paid are considered tourism by the US authorities and US citizens are prohibited from joining such vacations, as they are not deemed 'fully-hosted' events.

The Cuban Interests Section (2630 16th Street NW, Washington DC 20009; telephone: 202-797-8518) processes visa applications. We understand that visas take several weeks to be issued, should be applied for well in advance, and are granted comparatively rarely, except to business people, guests of the Cuban government, diplomats or other US government employees. Direct charter flights operate from Miami, Los Angeles and New York for those authorised to travel. Fuller details concerning the restrictions can be viewed at http://www.treas.gov/offices/enforcement/ofac/speeches/testimonycubatravel.pdf. A complete description of the embargo regulations can also be acquired from the Office of Foreign Assets Control website: http://www.treas.gov/ofac/.

We are aware that many US citizens travel to Cuba, despite the restrictions, arriving via Mexico, the Bahamas, Jamaica, the Dominican Republic or Canada. Information concerning flights to and from these countries into Cuba is presented at http://usacubatravel.com/aironly.htm. Travellers taking such a route have, in the past seemingly faced few risks, abetted by Cuban customs officials who stamp only tourist cards, not passports, on arrival and departure from the country. However, as is made clear in the most recent Treasury Department literature on the subject, the former Bush administration took a much harder line with those who flouted the ban on travel. It remains to be seen how the recently elected Obama administration will handle this issue, although there have been suggestions that the new government is interested in forging a new and more 'liberal' relationship with the Castro regime. Persons apprehended by US customs are advised not to lie, but should realise that they may be subject to a fine of up to US$7,500 by OFAC for breaching regulations. The issue of US citizens travel to Cuba is also addressed in depth by the American Association for the Advancement of Science's website, http://www.shr.aaas.org/rtt/.

*Currency and exchange rate*

Again, the only country where a special situation exists is **Cuba**. Until recently the country was effectively blessed (or cursed) with two economies – one peso and the other dollar – and three currencies, pesos, US dollars and convertible pesos (CUC). Since November 2004, US dollars have no longer been accepted as legal tender, and travellers are required to use convertible pesos, which should be purchased on arrival in Cuba. US dollars, British pounds and Euros can all be used to purchase convertible pesos, but note that dollars are subject to a 10% commission charge, which does not apply to pounds or Euros.

CUC possess no value outside Cuba, so ensure that you exchange any surplus currency prior to leaving the island.

Sterling travellers' cheques are valid in Cuba, provided they are not issued by American Express or issued by an American bank, but are not necessarily easy to change outside Havana. It is relatively straightforward to acquire a cash advance against most major credit cards in either a bank or an official CADECA exchange house (located at the international airport and many of the major hotels will also have one). In contrast, ATMs are still difficult to find and frequently break down; if you need to get money via this method it will, rather amazingly, be much less time-consuming to queue at a bank. Credit cards, especially Visa, are increasingly accepted for purchases in tourist areas, but it is always wise to carry a reasonable supply of cash. Importantly, it is impossible to use an American Express card under any circumstances, or indeed any credit card issued in the US or against a US bank.

While the average Cuban wage may be in the region of CUC15–20 per month (in 2008, the CUC had parity with the US dollar and sterling was worth 1.55 CUC), day-to-day costs in Cuba as a Westerner are nowhere near comparable. Cuba is competing within the West Indian marketplace for tourists, and the prices for services reflect this. As of 2008, hotels in the major cities and tourist regions generally cost at least CUC35 per night (including breakfast) for a double room, while the all-inclusive resorts on some of the cays cost at least double this amount. In addition, you should budget at least CUC10–15 for meals, snacks, drinks etc. It is advantageous, whenever possible, to carry a reasonable supply of small-denomination bills: waiting for what seemed like aeons while the waiter went to hunt for change, or a taxi driver had to drive around on the same mission were, at least formerly, not uncommon situations.

Visiting **Puerto Rico** in terms of money and costs is basically similar to the USA, so the currency is the US dollar and credit cards are very widely accepted.

On **Jamaica**, the Jamaican dollar is the official currency. Credit cards as well as bankcards are readily accepted. If you prefer to make purchases in small shops, or eat on the street, you should have a small amount of cash available, and it generally is not a good idea to carry large amounts of cash in Jamaica.

Most visitors to the **Caymans** are unlikely to stay more than a few days. US dollars are widely accepted, as are credit cards, making it probably unnecessary to purchase any local currency. On **Haiti**, the official currency is the Gourde, but the US dollar is frequently preferred. ATMs are available at the airport and large banks, and it is wise to rely on cash rather than using credit cards.

The currency in the **Dominican Republic** is the peso. Visitors should plan on acquiring sufficient local currency at the airport or in main towns, such as Santo Domingo, as credit cards are not necessarily widely accepted. Exchange facilities accept most major world currencies, but US dollars are perhaps most readily changed. If using a credit card in either the Dominican Republic or Cuba, note that some outlets (hotels and restaurants, especially) regularly add a surcharge of

5% or 11%, respectively, for so doing. In the Dominican Republic, as in all of the countries of the region, we would not advise visitors to exchange money on the black market.

***Travel insurance***

Comprehensive travel insurance is strongly recommended and should include full health cover and repatriation in case of a serious medical emergency.

***Time***

The different islands of the Greater Antilles lie within the same time zones as the eastern half of the United States, with Puerto Rico and the Dominican Republic being four hours behind Greenwich Mean Time, and Cuba, the Caymans, Jamaica and Haiti being five hours behind.

***Documents***

It is wise to carry your passport (and tourist card in those countries where these are necessary) with you at all times, as this document will be usually required to change money or when checking into a hotel. No health certificates are required, unless travelling from a yellow fever infected area. Carrying an international driving license is always useful, should you intend to hire a vehicle, but is not essential, even on Cuba.

# TRAVEL INFORMATION

Most travellers to the Greater Antilles will arrive by air, although some visitors, especially to the Caymans, Jamaica and Puerto Rico, might arrive by sea, as a short-duration stop on a Caribbean cruise. Anyone so doing can still do well for endemics on any of these islands. Each of the archipelagos in the Greater Antilles is served by an international airport, and sometimes more than one (in the cases of Cuba, Jamaica and the Dominican Republic). However, international flight options to and from Cuba, Haiti (especially) and the Caymans are comparatively limited, although not restrictively so, while transatlantic options to and from Puerto Rico are much fewer than to and from the mainland USA. It is worth noting that would-be visitors to Jamaica, Cuba and the Dominican Republic might find good package holiday options, with or without accommodation included according to island, at very competitive (and sometimes really bargain) prices. However, note that package trips to the Dominican Republic are usually based on the north or east coast, and therefore enter and exit the country via either Puerto Plata or Punta Cana, both of which are quite some distance from some of the main birding sites. It is definitely worth spending some time 'shopping around' for the best deals on the Internet. It is usually recommended to reconfirm your return flight between two and three days in advance, either by phone or online, although some airlines no longer seem to insist on this.

Those on 'hell-bent' birding trips with two or more weeks to spare can easily combine two (or more islands) in a single trip, especially if hiring local expertise to find the birds, though it is obviously dependent on flying between countries. For instance, in two weeks it should certainly be possible to see most or all of the endemics and specialities on Puerto Rico, the Dominican Republic and Cuba, or on Cuba, Jamaica and the Caymans, as just two possible combinations. For instance, GMF and GMK very successfully combined Cuba, the Dominican Republic and Puerto Rico into a trip of just over two weeks. A comprehensive trip through the entire region would be easily achievable if you have three weeks to a month at your disposal. Flights between most islands are frequent, e.g. between Cuba and Jamaica, Puerto Rico and the Dominican Republic, but slightly less so between, e.g. Cuba and the Dominican Republic, so some planning might be required. It is impossible to travel directly between Puerto Rico and Cuba, but it is also worth noting that as of 2009 it was also not possible to fly direct between Jamaica and either the Dominican Republic or Puerto Rico, with the only short-connection routes via Miami, in Florida. Grand Cayman is easily accessible from Cuba, making it surprising that very few visitors seem to take this opportunity to see Vitelline Warbler.

***Land transportation***

The general travel guides mentioned elsewhere in the introduction ably describe the options for travelling around any of the islands using public transport. We will not linger on these, except to mention that while many areas are served by buses (and the major cities on Cuba by trains), and some cities and tourist resorts, e.g. Cayo Coco on Cuba, by internal flights (best booked in advance through your travel agent or tour operator), hiring a car is by far the best option anywhere in the

Greater Antilles, except perhaps the Caymans where hiring a taxi should suffice. It might also prove possible to hitchhike, or to hire a bike or moped locally. Indeed, it may be perfectly feasible, if time-consuming and frustrating (e.g. remember that Cuban roads, compared to those in the West, are still virtually empty), to travel around many areas by a combination of hitchhiking and taxi. If you plan to travel by such means, remember that on Cuba and in the Dominican Republic you will probably acquire better deals and experience fewer hassles if you possess a reasonable comprehension of Spanish. However, wherever possible and especially if constrained by time, it is advisable to hire a car for the flexibility this provides.

A hire car is the best way of travelling around, although if you wish to visit the Oriente of Cuba it may be advisable to take a flight to either Holguín or, more likely, Santiago de Cuba (unfortunately the option of driving one-way and taking a flight in the opposite direction is usually either rather expensive or not even possible; only one or two companies currently offer such an option). Except on Cuba cars can be rented through the usual international and North American agencies, which have offices in most main cities, at airports and in some of the better hotels. You usually need to be over 21 years old to hire a car anywhere in the Greater Antilles. Anywhere in the region we would recommend that you book your car in advance, in order to get the most advantageous rates, but especially so on Cuba (see below).

Driving can be hazardous on Jamaica and in Dominican Republic, and due to the colonial legacy you drive on the left side of the road in the former country. Jamaican drivers can be reckless, and to a slightly lesser extent this comment and those that follow also apply in the Dominican Republic. Many vehicles do not have functioning tail or brake lights. There are few or no road shoulders and vehicles may stop on the pavement without warning, to pick up or drop off passengers. Many drivers do not use lights at night, and there can be livestock on the roads. Be very careful when driving and try to avoid driving at night.

On Cuba, where none of the well-known international agencies has an outlet, some of the better and more reliable of the local agencies include Micar, Havanautos (www.havanautos.cubaweb.cu), Transautos (www.transtur.cubaweb.cu), Cubacar (www.cubacar.cubaweb.cu), and Vía Rent-a-car (www.gaviota.cubaweb.cu). As elsewhere, prices vary, sometimes even between different agencies of the same company, and it may pay to shop around. We have experienced, sometimes considerable, difficulty renting a car at all when attempting to do so 'on spec' (even in Havana where all of the major agencies have multiple offices). Travellers should book in advance of their trip either through the hire agency direct, or via a reputable operator abroad. The high turnover in vehicle usage, especially for the smaller cars, may mean that some are not well maintained. Remember that costs are not especially cheap (which applies to anywhere in the Greater Antilles), although by pre-booking in sterling the price may be substantially better and that you will need a credit card to cover the refundable insurance deposit and the first tank of petrol when you collect the car. An international driver's license is, however, not necessary, just your regular documents and passport. You are not required to return the car with petrol in the tank. Remember to collect and destroy the security deposit voucher upon returning the car.

Also important to remember is that petrol stations are not what they are in the West, you are not always just around the corner from the next one! Even quite large towns may only possess one or two. Advance planning is always advisable, if in doubt fill-up when you can. Cupet owns most of the dollar petrol stations, but others (principally owned by Oro Negro) have started to open in recent years. Credit cards are becoming increasingly accepted at petrol stations in Cuba and elsewhere in the region. Like many things in Cuba, petrol is not especially cheap, even by Western standards, and most hire cars run much better using *especial* rather *normal* petrol. Despite the relative lack of traffic, few of Cuba's roads could be described as being in a high state of repair, thus flat tyres are not uncommon. Fortunately, most towns have a tyre repair (*ponchera*) facility.

# STAYING IN THE GREATER ANTILLES

*Accommodation*

We have usually refrained from making recommendations concerning hotels, or trying to provide up-to-date information on prices because such information changes so frequently. All of the recommended travel guides and tour operators can provide such information much more ably than ourselves. Lists of all the acceptable hotels that we could recommend in, for instance, Havana or Camagüey, on Cuba, or San Juan, on Puerto Rico, would only duplicate those mentioned in general tourist guides. Instead, we have limited our comments to noting the presence or absence of suitable accommodation within easy distance of a given locality, and to the relative merits of particular hotels in a tiny handful of cases, where we feel that this information may be of greater value to the visiting birder. However, it is worth noting that general hotel prices in the Greater Antilles average US$30 dollars per person per night, and that you are likely to spend quite a lot more than this on Puerto Rico and the Caymans, as well in the major tourist resorts on any island.

Lodging in most areas of Puerto Rico is relatively expensive, but there are some charming and reasonable lodgings known as paradors, which are state-sanctioned lodgings located in areas known for their natural beauty or historical interest. They are sprinkled throughout the country, and are among the least expensive accommodations available. For information on the paradors call 800-443-0266 (US) or +1-809-721-2884 from elsewhere in the world. There are no economical lodgings in Humacao, so don't waste your time trying to search one out. Just return to Fajardo or continue onward to Ponce or beyond.

On Cuba, some of the merits of the *casa particular* system, which has been authorised by the Cuban government in recent years, are described under the introduction to the Zapata region. There are certainly a number of benefits to their use (lower average costs being one), and birders being, in our experience, often adventurous, quite frequently do so. An introduction to this form of private enterprise, along with some other useful information for independent travellers to Cuba, can be found at http://www.cubatrip.com/english/index.html. Unless you plan to make use of such private houses, on Cuba there is considerable advantage to booking hotel accommodation in advance through a specialist operator, as you will generally acquire a much better deal than by just turning up.

*Food and drink*

When it comes to music and dancing, Cubans and Hispaniolans are born aristocrats, while they reputedly roll the best cigars known to man and Cubans own the finest set of American classic cars anyone is ever likely to witness, but gastronomically these islands are somewhat less famous. Typical fare consists of rice and black beans (*congrís*), pork, chicken or beef, eggs, bread, a slim selection of vegetables, salad (mainly cucumber and tomato, but also excellent, locally grown avocado), with pizzas, sandwiches and crackers serving for snacks. Everywhere throughout the Antilles, the quality improves in the better hotels (the old adage is certainly true, you do get what you pay for) and the variety of foods is strikingly better in most of the all-inclusive resort hotels on any of the islands. Because the quality of tap water on most of the islands is reasonable, stomach bugs acquired through a change of diet are somewhat less common than in parts of Asia for instance. Service is often rather slow, except in buffet-style restaurants, which are the norm in many hotels. Breakfast offers, to many, an irresistible opportunity to stock-up for a day in the field. There are also an ever-increasing number of reasonably cheap, fast food places (even on Cuba, which has the El Rápido chain). All of the well-known North American restaurant chains are present in abundance on Puerto Rico, meaning that if you are on a very short-stay trip you will not have to worry about losing time eating! Perhaps the most famous food on Jamaica is beef or chicken 'jerky'. Even on Cuba, vegetarians can survive but the choice can be relatively limited, especially outside of the main resorts and towns. GMK and GMF can recall a five-day stay in one hotel (which shall remain nameless) that steadily, and then dramatically, ran out of food; none of us has yet had the opportunity (or is it the courage?) to check whether that micro-scale situation has recently improved!

Also worth mentioning on Cuba is the availability of *paladares*, private houses that have been granted permission to set up as restaurants. These and *casas particulares* (which are explained in the opening section of the chapter dealing with Zapata) are perhaps the best option for those on a tighter budget, or who want to see more of 'real' Cuban life.

The Caribbean is the home of rum and the islands in general, perhaps especially Cuba, are famed for the inventiveness for their cocktails. One of the most famous is the daiquirí so beloved of Ernest Hemingway and less famously commemorated in the name of a fossil bird, *Siphonorhis daiquiri*, but for those who like a slightly less heavy tipple, the local beers are importantly cold and wet! Best known amongst the local beers are Cristal and Bucanero in Cuba, Red Stripe in Jamaica, Prestige in Haiti and Presidente in the Dominican Republic. The islands produce most of the finest rums of the world: Barbancourt in Haiti, Brugal in the Dominican Republic, Havana Club in Cuba, Myers and Appleton in Jamaica, and Bacardi and Captain Morgan in Puerto Rico.

Always carry sufficient drinking water, as the islands' climate is tropical and humidity levels can be very high. Bottled water is widely and cheaply available, as is a variety of packaged fruit juices and fizzy soft drinks, although cola addicts should note that on Cuba only the local brand is usually available.

***Language***

Three different languages are predominant throughout the region: Spanish on Cuba, Puerto Rico and in the Dominican Republic, English on the Caymans and Jamaica (the latter with a very heavy accent and very different slang), and French in Haiti. English, however, is also very widespread in Puerto Rico, and increasing numbers of people speak some English in both Cuba and the Dominican Republic. Nonetheless, birders are well known for venturing off the usual tourist trail, so in both these latter countries some knowledge of Spanish will prove to be of considerable assistance. In both Cuba and the Dominican Republic the Spanish is more Latin American than Castilian, and Cubans, in particular, are well known for speaking very fast, 'dropping' the ends of words and 'shortening' nouns. For non-native speakers, it's probably true to say that if you can learn Cuban Spanish then you can pick up any other variant of Spanish.

***Communications***

Except in Puerto Rico, phone calls home from anywhere in the region are likely to be either expensive or very expensive. Internet cafes are increasing everywhere and generally offer a far more cost-effective means of keeping in touch with friends back home or the football results! Only in Cuba are such services still generally very slow, because of the relatively poor internal telephone system, rather expensive to use, and often few and far between (at the time of writing, your best bet are the larger tourist hotels).

***Electricity***

Most sockets are of the American two-pin type (either flat or round), and the normal voltage is 110v to 125v, although some of the more modern hotels have 220v on Cuba. Most European visitors will probably need to carry transformers and adapters for charging batteries or other electronic equipment.

# CLIMATE AND CLOTHING

All of the islands of the Greater Antilles possess a fundamentally tropical climate, and all are arguably subject to as many, if not more, local variations in temperature than seasonal differences. Seasonal variation is most apparent in rainfall, rather than temperature shifts. On Cuba annual mean temperature ranges from 24°C to 27.6°C, with regional monthly mean differences of approximately 5°C and 7°C. Similar ranges are to be found on the Caymans and Hispaniola. The dominant wind direction is north-east, while relative humidity is 80% on Cuba and can reach almost 87% in some parts of the Dominican Republic. Drier regions may experience relatively lower average humidity, but this is still usually 70% or higher, though values below 40% are experienced, although rarely, on the Caymans.

On Cuba, temperatures can reach 40°C in summer (and as high as 43°C in the Dominican Republic) but have dipped as low as 1–2°C on Pica Turquino, the country's highest peak, in winter. Subzero temperatures have occasionally been recorded in the Dominican Republic, and may even be regular in the country's highest mountains. Such conditions are unlikely to be experienced on any of the other islands in the region, even in the highest areas of Jamaica. For instance the lowest recorded temperature on Grand Cayman is just over 11°C. The coolest months on Jamaica are likely to be December to February, but this is relative.

Anywhere in the Greater Antilles, the sunlight is frequently quite strong, so a cap is recommended, especially during the middle of the day. Sunscreen is also advisable and hiking boots are highly recommended, especially for montane trails. Although shorts and T-shirts are perfectly acceptable (though mosquitoes can be a real nuisance on several islands especially in the wet season), polypropylene or cotton trousers and shirts can be recommended. On Jamaica many areas are infested with hordes of seed ticks. It is wise to prepare your clothing in advance with permethrin (sold as Duranon).

It can rain at any season (so a small umbrella is always a good idea), but downpours rarely occur in the dry season (October–April in Cuba, November–April in Hispaniola, December–April on the Caymans). May and October are generally the wettest months on Jamaica. Regional annual rainfall ranges from 300mm to 7,500mm, with some of the areas of highest precipitation being Jamaica's John Crow Mountains, the Moa/Baracoa region of eastern Cuba, and parts of the Massif de la Hotte in Haiti. Ironically, one the driest, southern Guantánamo province, is also in eastern Cuba, and north-west Haiti also generally experiences very little rainfall, as do those parts of Hispaniola in the rainshadow of the higher mountains. Average annual rainfall scarcely exceeds 1,100mm anywhere on the Caymans. On Cuba and Jamaica, there is also a trend for rainfall to increase away from the coast. Given that it can be cool in montane regions, such as La Güira on Cuba, the Blue Mountains of Jamaica, and in the south-west Dominican Republic on winter mornings, cool-weather clothing is advisable. A light jacket will also be useful if travelling to Zapata between late November and late February; evenings and nights can be quite surprisingly and deceptively cool in this region at this time of year. Cold fronts from

North America can bring storms and lower temperatures to Cuba and the Caymans, much more rarely the other islands, between December and March.

Hurricanes pass through the Caribbean region almost annually in June–October, occasionally as late as November. Winds in these storms regularly reach almost 120km/h and up to 50cm of rain can fall in just 24 hours. The low-lying Cayman Islands can be extensively inundated by high seas during tropical storms, one of which, Hurricane Flora, brought winds of up to 320km/h to Hispaniola. Fortunately, the chances of witnessing one at any given location are very low, but most birders will want to avoid visiting the Greater Antilles at this season. In any case, during most of this period relative humidity is even higher, mosquitoes are more prevalent, and there are few migrants from the north to further entice the visiting birder (see When to Go). Should work or other circumstances leave you few options other than visiting in the tropical cyclone season, it should be noted that Cuba has a better than average weather report system in advance of such events, and has good experience in organising evacuations, so that casualties are minimal or non-existent.

Tropical storms can also be bad news for birds and other wildlife, especially when they strike forested areas, leaving large areas of trees fallen or dead and dying. Their effects have been most closely monitored in our region on Puerto Rico, especially on the endemic and globally threatened parrot, but may have a visible impact on other species too. For instance, on Cuba, in the aftermath of one recent tropical storm that made a direct hit on the Zapata Peninsula, Bee Hummingbird and Fernandina's Flicker were both distinctly harder to find for some time afterwards. This can make local knowledge very useful when planning your trip.

# GENERAL SAFETY, HEALTH AND MEDICAL FACILITIES

Neither Puerto Rico nor the Caymans can be regarded as dangerous countries to travel in, and while certain areas of Jamaica, such as downtown Kingston, are best avoided, this island too is generally very friendly. 'Caucasian' visitors may find it disconcerting to be occasionally called 'whiteys' to their faces, but no greater vigilance than one would usually exert is generally called for. It is a pleasure to report that Cuba must rank as one of the safest countries in Latin America, if not the world, for foreign travellers.

While the usual rules apply, and you should always remain alert and watchful, you will be extremely unfortunate to be the victim of a violent crime, which carries high penalties for convicted offenders. Pick-pocketing and bag or camera snatching, while apparently on the increase, are still comparatively rare events, and also subject to severe penalties. At most, you are likely to be subject to the unwanted attentions of touts eager to sell you cigars, rum, a place to stay, etc.; seasoned travellers will have no difficulty in dealing politely but firmly with such people. And, quite often, the same people who tried to sell you something you didn't want will prove just as willing to simply talk for the novelty of conversing with a Westerner, especially if you speak some Spanish or they English.

Cubans are a spectacularly friendly race, and in some ways a breed apart from the rest of humanity. Those with an open mind and a willingness to go with the flow (remember this is Latin America, where life automatically moves at a slower pace) will surely enjoy every minute of a journey through Cuba. Dominican Republic too is largely safe, though one should generally steer clear of particularly poor areas. Anyone intending to visit Haiti, however, should exercise extreme caution, especially in the environs of the capital, Port-au-Prince, as there is a certain degree of lawlessness and kidnappings have become relatively frequent in recent times. When this book was in proof, Haiti was tragically hit by a devastating earthquake that left over 200,000 people dead, and destroyed much of the country's already spartan infrastructure. The region between Port-au-Prince and Jacmel was the most badly affected.

There are no specific vaccinations required for entry to any of the Greater Antilles, although those arriving from or continuing to South America may be asked to present a yellow fever vaccination certificate. Proof of a cholera vaccination may also be demanded if travelling from an area that has recently suffered an outbreak. It is advisable to ensure that your vaccinations against typhoid, tetanus, hepatitis C and polio are up to date. Malaria has been eradicated in the region, except in Haiti and the Dominican Republic. However, dengue is still a risk throughout the Greater Antilles.

Medical care is no longer free in many parts of Cuba for foreigners, but still reasonably good, despite many hospitals' rather spartan appearance, and can be swift, even for comparatively minor ailments. Mike Flieg had to visit a hospital as the result of a fall from a cliff, which resulted in a severe laceration. Not only was the wound stitched (with bright blue fabric thread), but he received follow-up care at the hotel. The hospital refused to accept compensation. Remember that many medicines can be to all intents and purposes practically impossible to obtain locally, even given the necessary money, so ensure that you bring sufficient supplies of any prescribed drug you take regularly. Even such apparently mundane items as painkillers for headaches and the like could be problematic to obtain locally, so prepare in advance.

Medical facilities on other islands in the region are generally good and inexpensive, to excellent but more expensive on Puerto Rico and the Caymans, so ensure that your travel insurance is up to date. Most people are unlikely to suffer anything more severe than an upset stomach anywhere in the Greater Antilles, and the risk of this will be minimised by drinking bottled water (though tap water in Puerto Rico at least is of reasonable quality). Sunstroke is the other principal risk; drink plenty of water and wear a hat, as well as applying sufficient sunscreen.

There are several mildly poisonous plants in the Greater Antilles, but you are unlikely to have any problems, especially if you keep to trails. The commonest belongs to the genus *Comocladia*. The plant is easily recognisable by its compound, bright green, saw-like leaves. No dangerous, poisonous animals inhabit the Greater Antilles. The black widow spider and several scorpions are rather rare and spend daylight hours out of sight. However, sleeping outdoors anywhere on these islands should be generally avoided. Try to avoid direct contact with centipedes and wasps, all of which have a painful sting.

Anywhere in the region during the summer and part of the rainy season (May–September), mosquitoes and sand flies can be very unpleasant, especially in Cuba's Zapata Swamp and on the cays. In November–April, their numbers are tolerable, and in some places they are absent. A good repellent is virtually essential, and wearing long-sleeved clothing is advisable at any season. It might be advisable to spray your clothing with Duranon before entering the bush, as seed ticks can be numerous and relentless. The bites are reminiscent of chiggers and will irritate for many days.

# BOOKS AND MAPS

Like so many areas of the world, birders are now well served by a range of books, audio guides, general travel guides, and maps to the Greater Antilles. Just ten years ago, this was far from true; indeed, the transformation has been extraordinary. Coupled with the explosion of up-to-the minute (or at least month) birding information available on the internet, birders have virtually no excuse for not being extremely well 'armed' on a trip to the region covered by this book.

***Field guides and other works***

Bond's classic field guide to West Indian birds, for so long the only available source for those seeking to identify Antillean birds, has been dramatically superseded only in recent years. Essential for anyone travelling to the region covered by this book is Herb Raffaele *et al.*'s *Birds of the West Indies* (1998, published by Princeton University Press/Christopher Helm). In 2003 a very handy field guide version of the original hardback book was published with a soft cover. The text has been slimmed down to the 'bare essentials', but improved illustrations were provided for quite a number of species, and colour maps for all. A field guide solely devoted to Cuban birds was published recently; it contains detailed text and illustrations of the endemics and all other species known to occur in the archipelago (Garrido & Kirkconnell, 2000, published by Cornell University Press/Christopher Helm). Those visiting Hispaniola are now extremely well served by the even more recently published *Field Guide to the Birds of the Dominican Republic and Haiti* (Latta *et al*., 2006, Princeton University Press/Christopher Helm). A great many of the illustrations are 'borrowed' from the West Indies field guide, but the endemics and some others have been repainted, and the text is brand new and extremely detailed.

Also potentially useful, as it includes many of the Greater Antillean endemics, is *A Photographic Guide to Birds of the West Indies* (Flieg and Sander, 2000, published by New Holland). Photographic works to Jamaica and Puerto Rico have been published, the latter also containing a CD with sounds and many more images than are contained in the book (see the Bibliography for details). A new book for Jamaica has very recently been published and offers a reasonably comprehensive selection of photos of the island's birds, including all of the regularly occurring species, as well as the most accurate and up-to-date distribution maps.

Several field guides to North American birds may prove of additional service for identifying Nearctic migrants (and even vagrants). Especially informative and well illustrated are *The North American Bird Guide* (by David Sibley, 2000, published by Houghton Mifflin/Christopher Helm) or the National Geographic Society's *Field Guide to the Birds of North America* (published by National Geographic, and comprehensively revised most recently in 1999). All of these guides are available from specialist retailers such as, in the UK, the Natural History Book Service (www.nhbs.com), Subbuteo Books (www.wildlifebooks.com), and WildSounds (www.wildsounds.com), and, in the USA, the American Birding Association Sales (www.americanbirding.org), Buteo Books (www.buteobooks.com) and Los Angeles Audubon Society (e-mail: laas@IX.net-com.com).

Among reference works, of interest to those attempting to contextualise their observations or intending to visit any of the islands more than once, both Hispaniola and the Cayman Islands are well covered by recently published checklists in the British Ornithologists' Union Checklist series. Further details on both can be found in the Bibliography. Both are highly recommended sourcebooks on their respective archipelagos. A detailed résumé of the status, distribution and habitats of Cuban birds is also currently in preparation; this will replace the now severely outdated *Catálogo de las Aves de Cuba* (Garrido and García Montaña, published in 1975).

For information on globally threatened birds, the BirdLife International website (www.birdlife.org) is an extremely useful source of data. Follow the links to the datazone, which allows you to view the latest information on any species currently considered globally threatened. In the same vein, the same organisation has recently published an extremely attractive and useful Important Bird Areas inventory for the region (in late 2008).

Those with an interest in other fauna and flora may wish to invest in some of the following guides covering other groups: *Butterflies of the Caribbean and Florida* (Stiling, 1999), the *Collins Pocket Guide to the Coral Reef Fishes of the Indo-Pacific and Caribbean* (Lieske and Myers, 2001), the *Peterson Field Guide to Southeastern and Caribbean Seashores* (Kaplan, 1999; covers plants, crabs, corals and other groups) and *Flowers of the Caribbean* (Lennox and Seddon, 1990; treats only common species). Anyone wishing to identify whales and dolphins that occur in Caribbean waters are advised to seek out the recently published *Whales, Dolphins and Seals: A Field Guide to the Marine Mammals of the World* (Shirihai, 2006, published by A. & C. Black). All of these are currently available from the Natural History Book Service.

***Audio Guides***

If you are serious about finding all (or at least many) of the endemics or have a general interest in vocalisations, you will also want to bring a tape-recorder, mini-disc or DAT (digital analogue tape) recorder. As long ago as 1978, George Reynard and Orlando Garrido produced a two-record LP entitled *Bird Songs in Cuba*, published by the Cornell Laboratory of Ornithology, which contained the songs and calls of 123 species and should still prove very useful in identifying some of the rarer birds that are more often heard than seen. (However, do note that the calls attributed on the record to Giant Kingbird actually refer to Loggerhead Kingbird.) Cornell has recently published the LP on CD, and the same publisher has also issued a comprehensive sound guide to Jamaican birds, as well as a brand-new set of CDs to birds of the eastern Caribbean (from Puerto Rico to Grenada) as well as covering the Caymans, Bahamas and the Colombian-owned island of San Andrés. Bear in mind too the Puerto Rico photographic guide, produced by Mark Oberle, which also contains sounds of many of the island's species. A number of the family audio guides, e.g. those covering New World owls, New World rails, vireos, etc., and published by ARA Records (Florida) or the Cornell Laboratory of Ornithology (New York), offer useful resources for those interested in Caribbean bird vocalisations. Among the best stockists of such guides are Bird Sounds (www.birdsounds.nl) and WildSounds (see address under Field Guides and other works). For those based in North America, and seeking only Cornell publications, it may be easier and quicker to purchase from them direct (www.birds.cornell.edu or www.birdsource.org). Also of invaluable assistance in this respect are the web pages of www.xeno-canto.org, a searchable archive of Latin American bird sounds, including all of the islands covered in this book.

***Trip reports***

The latest information likely to be useful to a birder, such as up-to-date news on travel, accommodation and, of course, bird sightings is liable to be found in trip reports either privately circulated or, these days, freely available on any number of internet sites. Most birders probably have their favourite web pages for accessing recent trip report information, but those hosted on Surfbirds (www.surfbirds.com) and Travelling Birder (www.travellingbirder.com) are regularly among the best.

***Travel guides***

Along with virtually every other part of the world, the Greater Antilles are now well covered by a selection of general travel guides, especially to Cuba, Hispaniola and Jamaica. Indeed, during the last few years the choice has become ever broader. However, in our opinion, a small number stand out. If you are intending to visit more than one island, then to save weight consider either *Caribbean Islands* (Miller *et al.*, published by Lonely Planet, and updated in 2008) or *The Caribbean* (published by Rough Guides, and updated in 2005). Do bear in mind, however, that both these books cover a far larger region than just the Greater Antilles, so they might not provide quite the level of detail you are searching for.

For single-island use, recommended are the *Cuba Handbook* (Cameron, published by Footprint Handbooks and updated, most recently, in 2002, but perhaps no longer in print), the *Lonely Planet Travel Survival Kit* (Sainsbury, published by Lonely Planet; most recent update 2006) and *The Rough Guide to Cuba* (McAuslan and Norman, published by Rough Guides; most recent update in 2007). Although we have not seen a recent edition, the *Traveller's Survival Kit: Cuba* (Calder and Hatchwell, 1996, published by Vacation Work) has also served us well in the past. Recommended for snorkelling enthusiasts is the *Diving and Snorkelling Guide to Cuba* (Williams, 1999, published by Pisces Books). For Hispaniola, outstanding is Lonely Planet's *Dominican Republic and Haiti* (Chandler and Chandler, most recently updated in 2008) or Rough Guides' *Dominican Republic* (Harvey, 2005), and both these publishers also issue books to Jamaica (Lonely Planet, 2008, and Rough Guides, 2006). Lonely Planet also publishes a single-country guide to Puerto Rico (Sainsbury and Cavalieri, 2008). Most or all of these should be available from any reasonable bookshop, and many can be obtained from the specialist ornithological retailers mentioned in the previous section, as well as online direct from their publishers (e.g. www.lonelyplanet.com and www.roughguides.com).

***Maps***

Travel in the Cayman Islands is generally unproblematic in this respect. Free maps are available, e.g. at the international airport, and are very good, being sponsored by firms such as AT&T and American Express. Since the islands have few roads they are more than adequate.

For Jamaica, we can recommend the either the Shell or Esso-sponsored road maps, both of which are very good, offering inset maps of the main towns and cities and an overall scale of 1: 356,000. They can be purchased locally or through good map suppliers abroad. The Globetrotter 1: 300,000 map is also highly recommended, and appears to be more detailed especially for the Blue Mountains.

Visitors to Puerto Rico will find that perfectly adequate maps are available locally, e.g. at the international airport, and that the local road system is very easy to navigate. However, if you do prefer to purchase a road map in advance of travel then we can recommend the very detailed map published by International Travel Maps, of Vancouver, Canada, with a scale of 1: 190,000. It has served us well in the past. Also recently recommended is the MetroData road map (which can be purchased online at: www.metropr.com).

A variety of maps are available covering Hispaniola, with the accent on the Dominican Republic. However, of those produced by the major publishers and widely available abroad, the best we have seen is the 1: 500,000 Marco Polo / Shell ReiseKarte map, although it only covers the easternmost part of Haiti, and does not include most of the few birding sites in the latter country. Those only planning to visit the Dominican Republic should find it more than adequate for their needs.

For Cuba, maps are frequently rather basic, and generally speaking not particularly useful. Although maps can be purchased in some bookstores and tourists shops on the island, those planning their own itinerary using a hire car will appreciate better and more up-to-date information, as road signs in Cuba are not among the best we have seen! Until recently there were few good maps of Cuba available outside the country. Things have changed and the independent traveller needing to rely on such information can now purchase several perfectly serviceable maps. Among the best we have seen is that published by Nelles, which has inset street maps for Camagüey, Cienfuegos, Havana, Santiago de Cuba, Trinidad and Varadero. Also recommended, by others, are the Freytag & Berndt, and Rough Guide maps of the island. Both should be widely available in bookshops.

If you can wait until your arrival in Cuba and have sufficient time, visit the Instituto Hidrográfico, in Old Havana, in Mercaderes, between Obispo and Oficios, which will hopefully have in stock either or all of the following: Mapa Geográfico (a large map of the island with smaller regional maps), *Guía de Carreteras* (recommended as the best map for drivers, published by the Cuban Directorate of Tourism in cooperation with Havanatur) or, the slightly less useful, *Cuba, Mapa de Carreteras* (though this does have a useful inset of the Havana region). There are also several good provincial maps but these are rarely obtainable outside of the region in question. If you are self-driving, purchase of the *Guía de Carreteras* will undoubtedly make travelling around the country much easier. It may be possible to purchase it in advance by e-mailing percorsi@ip.etecsa.cu or azludovici@libero.it. You should also check online (www.cubamaps.com or www.cubadirecto.com). Nonetheless, we should point out that most drivers in Cuba will probably get by with a foreign-purchased map and some knowledge of Spanish. Many birders carry a GPS system these days, but it should be noted that we have heard reports of such equipment being confiscated by Cuban customs officials, who may even prove suspicious of more routine birders' kit, such as telescopes, binoculars and tape-recorders.

Some of the retailers of general ornithological literature, such as the Natural History Book Service or Subbuteo Books (for their addresses see Field Guides and other works), often stock a range of maps useful to the visiting birder. In the UK, specialist map and travel guide retailers, such as Stanfords (www.stanfords.co.uk) can also be recommended.

# WHEN TO GO



As the primary reason for most birders to visit any island in the Greater Antilles will be to find the endemic species, then it should be stated at the outset that this is feasible at virtually any season of the year. However, most of the specialities are more easily found during the breeding season (generally March–July in Cuba and Hispaniola, and similar for other islands). For transients, perhaps including a vagrant, the best months are August (especially the first two weeks), September (the second and third weeks), October (especially the second and third weeks), and March or April (especially the first two weeks). The best time to see a good selection of winter visitors is between October and March, when the overall numbers of species present on the islands is highest. The small number of migrant breeders generally start to arrive in March and April, e.g. Antillean Nighthawk, but note that Cuban Martin arrives on Cuba from the end of January, although it is not generally very widespread until March. Given the hot, often humid (and sometimes very wet) conditions that prevail in summer, most birders, especially those from Europe, will find that a visit anytime between late November and late April is liable to be most productive. If observing a wide range and the largest numbers of North American migrants, especially warblers, is a priority then it is best to visit the islands before the end of February, otherwise seeing large numbers of individuals will be to some extent dependent on experiencing a 'fall-out'.

As in many areas of the world, the Caribbean is a popular part of the world to visit at either Christmas or Easter, thus airfares and hotel prices are frequently inflated at these times. If possible they are best avoided or, if you must visit during one of these periods, then be sure to book both your flights and accommodation as far in advance as possible. Hurricanes pass through the Caribbean region almost annually in June–September, and this season of the year is generally best avoided. Nonetheless, hurricanes and their paths are predicted and plotted weeks ahead, making a quick visit to the region possible at any time.

# INTRODUCTION TO THE SITE INFORMATION

The bulk of the book is, of course, the locality accounts. For ease of use, we have divided the Greater Antilles into its islands and countries (note that the Cayman Islands are British territory and Puerto Rico a US commonwealth territory). Thereafter, where relevant, we subdivide each island or country into different geographical areas. For Cuba, we commence with the Zapata Peninsula, which is arguably the most famous birding locale in the entire Caribbean region. Any birder heading to Cuba is strongly advised to spend a few days in the area, even if this necessitates missing other localities. Put simply, if you only have a few days for birding on Cuba, all roads should lead to Zapata.

Each site is generally covered by an introductory text. We have rarely mentioned specific hotels / restaurants and their relative merits / costs due to the propensity for such information to change, and because most are already well covered by many of the general tourist guidebooks. Because public transport is generally inadequate for reaching the main birding sites virtually anywhere in the region, especially Cuba, we have assumed that any birders using this guide are likely to have their own transport. Remember that distances, which are quoted in kilometres, are liable to be accurate only as much as the odometer used to measure them, so don't be surprised if the readings on the odometer in your hire vehicle don't always match those presented here.

Throughout this book we focus the vast majority of our attention on sites that will be visited time and time again by birdwatchers in search of single-island or regional endemics. Several sites are covered in only a few lines; these are sites that we have visited only very rarely (very occasionally not at all) and are likely to be of lesser interest to most visiting birders. Nonetheless, such areas have been included because, for instance, we consider that birders holidaying with their families in one or other of the hotels that have sprung up on some of Cuba's offshore islands in recent years will appreciate knowing that binoculars should still be packed! We do not usually provide detailed site information for these areas, as usually there are no specifics; any area of cover or suitable shoreline is likely to hold birds.

English names and taxonomy follow those used in Herb Raffaele *et al.*'s *Birds of the West Indies* (Princeton University Press / Christopher Helm, 2003), as this is the only field guide in print covering the entire region. Bird lists for any given site always focus on endemics and near-endemics (even where these might be quite common and widespread), along with other specialities (for instance globally threatened species).

# CUBA



A very brief potted history of the country's ornithology, as opposed to birdwatching as a leisure activity, was presented in the *Field Guide to the Birds of Cuba*. Pioneers, in the late 1950s, of birding for fun were the late Florentino García and Domingo Fernández Montaner. Due to the former's knowledge of some good areas, and through his membership of the Audubon Society, they planned to organise tours to several areas near Havana, such as Jibacoa and Soroa, but their plans were thwarted by political events. About 20 years later, George Harrison, a staff member of the International Wildlife Organization, visited Cuba to report on the island's birdlife and to explore the possibilities for birdwatching. In 1979, he produced a fascinating article, with superb photographs, sparking the minds of aficionados. Besides detailing his experiences, he placed special emphasis on a section entitled 'How to go birding in Cuba'.

***Map of Cuba***

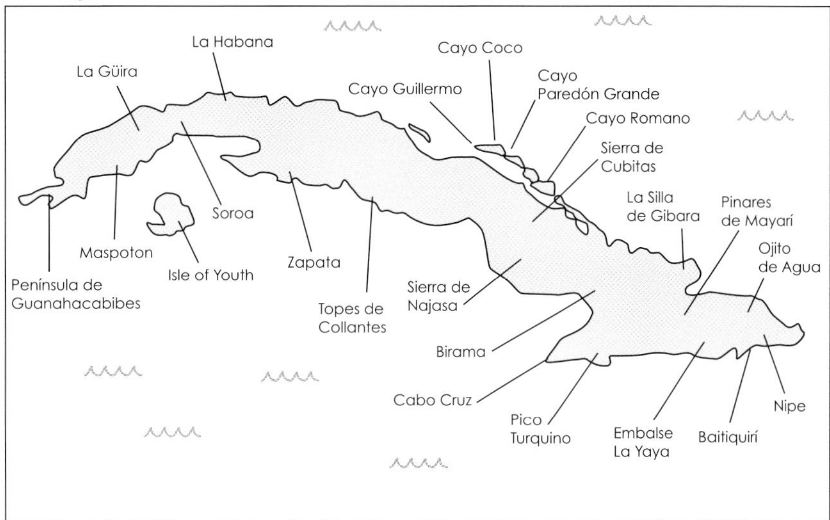


Soon after the article was published, a group of 16 keen North American birdwatchers arrived in Cuba. They were headed by the late James F. Clements, author of a *Checklist of the Birds of the World* but despite the trip's undoubted success, it was not followed by others, mainly due to the political situation.

Around 1985, two well-known Canadian writers, the world-renowned novelist Margaret Attwood and her husband, Graeme Gibson, visited Cuba through their literary interests. As they were also keen birdwatchers, they managed to take half a day to visit Zapata Swamp. Gibson was fascinated: so much so, in fact, that soon after he started promoting birding tours to Cuba, departing from Toronto on charter flights.

In addition to such organised trips, there were occasional visits by small groups of birdwatchers from other countries, especially Germany and England. Of these private travellers, Marcus & Simon Sulley wrote a report on their visit, entitled *Birding in Cuba*. Most of the Canadian and American groups were led in their early years by Orlando Garrido, and his dear friend and field colleague, Rogelio García Arencibia ('Pelao'), who by then was 70 years old. Rogelio was born in a small village in

Zapata, called Soplillar. For 11 years they guided tourists and naturalists into the swamp, prior to Pelao's death in January 1991. In January 1988, Arturo Kirkconnell joined this team. In recent years, several local birding guides have taken Rogelio's place in the swamp, of which the most outstanding are Osmany Gonzalez (who has lived in the US since 2006), and Orestes Martinez (known simply as 'El Chino').

Subsequently, in the early 1990s, the first of the specialist birding tour companies from Europe started organising trips to Cuba, and a little later, in the latter part of the same decade, interest in Cuba began to snowball among Western birders. Currently, at least ten birding tour operators organise trips to the island on an annual basis. At present, there are two bird-tour leaders in Cuba, William Suárez and Arturo Kirkconnell, both of whom work at the National Museum of Natural History in Havana.

# Zapata Peninsula

Located in southern Matanzas Province, just over 150km from Havana (2.5 hours drive), the peninsula occupies an area of 4,700km$^2$ and supports several different natural habitats, including semi-deciduous forest, swampy forest, mangrove, coastal thickets, dry forest, mudflats, and extensive wetlands, including saltpans and fresh and brackish marshes. A dense layer of organic mud covers almost 70% of the region; the drier sections of Zapata have abundant limestone and rather thin soils. The entire area is very low-lying, with a maximum elevation of just 5m!

*Map of Zapata Peninsula*

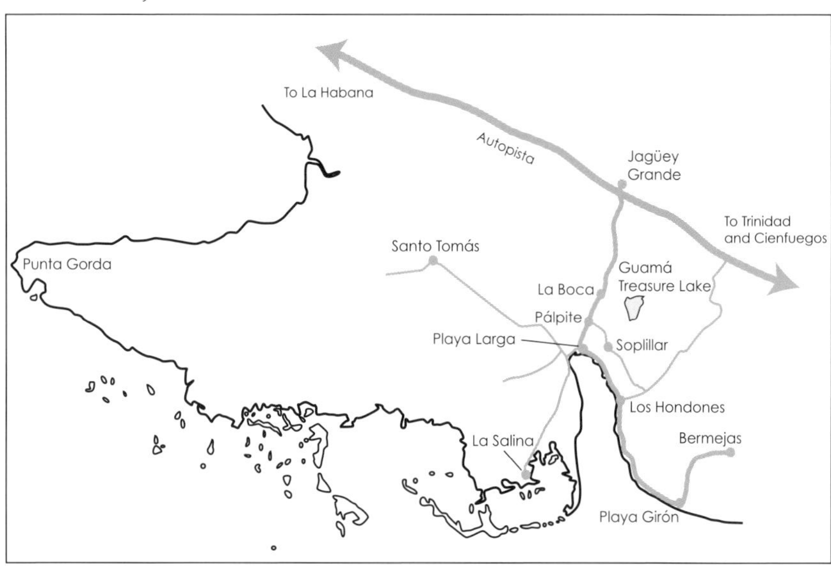


The journey to Playa Larga from Havana takes about 2.5 hours. At km142 on the freeway between Havana and the east of the island, turn right (south) towards Playa Larga (the turn is clearly signed to Zapata), then shortly afterwards take the right turn in front of the large factory at the small town of Australia, and the resort is 29km along this road.

Most birders will prefer to stay in one of the two comfortable, but relatively basic, tourist hotels in the region, Playa Larga, or the more distant Playa Girón, which is 32km further east along the coast. Both possess similar facilities, including a swimming pool and easy access to the beach, and all of the sites listed in this chapter can be comfortably visited within a day from either. Indeed, at most times of year it is sufficiently hot (and bird activity insufficiently high) to make returning to hotel for lunch and a short siesta an attractive and feasible option. If you prefer to remain all day in the field, stock-up with provisions at the buffet breakfasts or purchase some snacks at one of the resort's on-site shops.

Accommodation at both resorts is in small, but reasonably comfortable, en suite private chalets. Playa Larga is better positioned for easy access to most of the main birding sites in the Zapata region, and is especially convenient for visiting either Santo Tomás, or the two sites in the northern part of the swamp – Peralta and Hato de Jicarita – as a pre-dawn start is essential for visiting all of these localities. Nonetheless, Playa Girón still makes an excellent base, and has the advantage of being positioned close to the especially endemic-rich sites of Bermejas and Punta Perdiz. Both Playa Larga and Playa Girón offer reasonable birding on-site, and both resorts possess an easily seen pair of Stygian Owls within their grounds. Another, much less convenient, but scenically very attractive, option is to stay at Guamá, within Treasure Lake, which offers many excellent birding opportunities, but is distinctly less ideal in several other respects, especially if you intend to do much night birding, as all transportation to and from the hotel is by boat. In addition, in our opinion, the on-site facilities and restaurant are not quite of the same standard as those at the other two resorts mentioned above.

Finally, there is the option to stay in private accommodation, in government licensed guesthouses (private homes), in Playa Girón, Playa Larga or even Jagüey Grande, although we would not recommend the latter, except for its close proximity to Peralta and Hato de Jicarita. Such accommodation usually costs $15–25 per person per night, while meals can usually be obtained for a few dollars extra. Most such accommodation is clearly indicated, but some of those in Playa Larga are best found by asking for directions from local people. (In recent years, several birders have strongly recommended a *casa particular* in Playa Larga run by Nivaldo Ortega Gutiérrez but there are quite a number of others.) Staying with a Cuban family in this manner provides some insight into local life, but is not necessarily a lot cheaper than staying in one of the hotels. Nonetheless, it probably offers greater flexibility and most travellers to Cuba who try to break the habit of staying in hotels enjoy the experience. It should be made especially clear that if using such *casas particulares*, you should only use those that are licensed by the government. The owners of licensed establishments will always record details of your passport in the official registration book, and should have a government sticker on their front door. Cubans caught renting rooms to foreigners without the necessary government authorisation are liable to be heavily fined, and you may come under considerable pressure to assist them financially as a result.

We suggest that most visitors should consider spending a minimum of four or five days in the Zapata region. In the following pages we present a near-exhaustive list of localities liable to be of interest to birders, but we suggest those with limited time should concentrate on the following, most important areas: at least one of Santo Tomás, Turba, Peralta or Hato de Jicarita, as well as Los Sábalos, Bermejas (which in the last few years has been the single best site for most of the woodland birds), Pálpite, Mera, El Cenote and Los Canales. These areas should offer the keen birder the opportunity to find most of the endemics and specialties of the region.

***Other wildlife***

In addition to birds, among other endemic fauna perhaps the most impressive is the Cuban Crocodile *Crocodylus rhombifer*. The likelihood of finding this species in the wild is very low, with the best locality being the Río Hatiguanico, near Hato de Jicarita, although the only sure place to see one is at the Boca crocodile farm, not in the wild. American Crocodile *C. acutus* also occurs. The endemic Cuban ground lizard *Leiocephalus cubensis*, two unique subspecies of the Cuban giant anole *Anolis luteogularis*, as well as a subspecies of another endemic lizard, *Leiocephalus stictigaster*, all occur in Zapata. The peninsula harbours a further four endemic species of anoles, three endemic snakes, including the Cuban racer *Alsophis cantherigerus* and Cuba's giant boa *Epicrates angulifer*.

There are also a few native mammals. Several species of bats spend the day resting in hollow trees and caves. Hutias are comparatively large, largely tree-dwelling rodents with incredibly wiry fur slightly reminiscent of that of a beaver. Three species inhabit Zapata, the two largest are fairly common, though not necessarily easily encountered, but the smallest, Dwarf Hutia *Mesocapromys nanus*, is extremely rare and on the verge of extinction.

During March and April, the road between Playa Larga and Playa Girón can be virtually covered with crabs *Geocarcinus ruricola* migrating from inland forests to the sea, where they lay their eggs. They are variously coloured red, black and pale cream. So abundant are these creatures that nine out of ten rustlings that you hear on the forest floor can be traced to a crab.

***Birds***

The Zapata Peninsula is, undoubtedly, the best birdwatching area in Cuba, and possibly the entire Caribbean region. It supports all but four of Cuba's avian endemics, as well as many other native species, both winter residents and transients, along with several summer and spring visitors (which breed in Cuba but return south in autumn). Almost 260 species have been reported in the area, which was declared a UNESCO Biosphere Reserve in the year 2000. We have selected a number of different locations for inclusion here, based on several criteria, including relatively high diversities of birds, easy access, and greatest variety of habitats.

## Playa Larga

This resort is perhaps that most commonly used by birdwatchers staying in Zapata. The complex consists of four streets with chalets on each side, a restaurant, bar, nightclub, and several small souvenir shops. The restaurant is right at the beach. The closest town is Jagüey Grande, about 30km to the north.

***Location*** Access is straightforward from the main freeway between Havana and the east of the island. Turn south at the crossroads at km142 on this road, just beyond a Rumbos bar with a boat beside it, which is on the same side of the road as the exit. After about 1km take a right-hand turn (the first one along this road) just ahead of a large citrus processing factory. Continue south along the road, passing the Boca resort and through Pálpite village, to Playa Larga village where the road bends to the left. The entrance to the hotel is on the right, a couple of hundred metres further along the road.

***Strategy*** There are several tall trees in Playa Larga resort, and the forest is very close on the opposite of the main road. Coconut trees may harbour Yellow-bellied Sapsucker and Yellow-throated Warbler in winter and spring. The most important feature of the Playa Larga avifauna for visitors is the pair of Stygian Owls that can be reliably found in the area. They are best searched for on dark nights, when the moon is either new or in its first quarter, at which times the male may be heard calling from several areas within the compound, although the isolated large tree behind the swimming pool and those tall trees close to the restaurant are consistently favoured. Later in the lunar cycle, when the nights are not so dark, the birds may be very elusive or even impossible to find; the same holds true for the pair at Playa Girón. It also pays to stroll along the shore among the sea grape trees, where you may find Northern Flicker, and several species of shorebirds. Situated at the head of the Bay of Pigs, Playa Larga does very occasionally attract seabirds, such as terns and gulls, including, in 1999, Cuba's second Franklin's Gull, and most recently, in spring 2009, a small flock of Black-legged Kittiwakes. Less than 1km from the cabins, along the coast towards Playa Girón, is a huge *Ficus* tree. Its blossoms may attract several species, including Red-legged Honeycreeper, which, in Cuba, is most easily seen in Pinar del Río Province. Bee Hummingbird has occasionally been observed at these flowers in the past.

***Birds*** Perhaps surprisingly, as many as 115 species have been recorded in this general area. Within and just outside the resort look for Yellow-crowned Night Heron, Cuban Black Hawk, Zenaida and White-winged Doves, Rose-throated Parrot, Great Lizard Cuckoo, Belted Kingfisher, Antillean Palm Swift, Cuban Emerald, Cuban Pygmy Owl, West Indian Woodpecker, Northern Flicker, Grey Kingbird (summer), Loggerhead Kingbird, Cuban Martin (summer), Cave Swallow, Black-whiskered Vireo (from mid-March), Cape May Warbler (winter and spring), Cuban Crow (low density, but obvious by its 'parrot-like' calls), Greater Antillean Grackle, Cuban and Tawny-shouldered Blackbirds, and Cuban Oriole.

## El Roble

*Location and Strategy*

This is one of the closest birdwatching areas to Playa Larga, just 7km south-east of the resort by the road towards Playa Girón, and situated directly opposite a camping area gate. The main vegetation types in this region are semi-deciduous forest, evergreen forest and second growth. It is best to walk about 200m along the trail until you come to an area of calabash or Güira trees (*Crescentia cujete*), which produce the very hard fruits used to make the well-known maracas (a Caribbean musical instrument). Its seeds are highly preferred by ground-feeding doves, with luck including Blue-headed and Grey-headed Quail-Doves. Beyond the trees is a glade, which is a very good area to observe a number of other species, especially around a grassy depression that becomes flooded during the rainy season.

*Birds*

Along this trail, the following species – and many others – have been observed: Key West and Blue-headed Quail-Doves, Rose-throated Parrot, Great Lizard Cuckoo, Cuban Emerald, Cuban Tody, Cuban Green Woodpecker, Crescent-eyed Pewee, Loggerhead Kingbird, Red-legged Thrush, Northern Waterthrush, Ovenbird, American Redstart, and Prairie, Black-throated Blue, Yellow-throated and Yellow-headed Warblers. This was formerly also a good area to find the uncommon and secretive Swainson's Warbler.

## Soplillar

This small village, about 5km north-east of Playa Larga, is the gateway to a number of interesting sites in the Zapata region. Drive south-east from the resort, towards Playa Girón, and take the first paved road on the left, which accesses Soplillar.

## Mera

Approximately 100m south-east of Soplillar (see above for directions to this village), take a narrow road on the left (just before a cattle grid), and drive about 150m until you reach a fork. Walk the trail on the right, which initially passes through a gate and then secondary forest, with chances for Blue-headed and Grey-headed Quail-Doves, Yellow-throated and White-eyed Vireos (both winter), Cuban Vireo, Blue-winged, Worm-eating and Palm Warblers, American Redstart, and Yellow-faced Grassquit. Once you reach an open area with scattered palm trees, patches of swampy forest and semi-deciduous woodland, where the ground can be very wet, you also have a chance of Cuban Parakeet, Rose-throated Parrot, Cuban Pygmy and Bare-legged Owls, and Bee Hummingbird. In addition, this is one of the best areas to find Gundlach's Hawk and Fernandina's Flicker during the breeding season. The latter is best searched for among the trees surrounding the final section of trail before the open area. You should also find West Indian Woodpecker and Cuban Blackbird. When returning, look for Northern Caracara among the high and scattered *Ceiba* trees.

*Map of Mera*

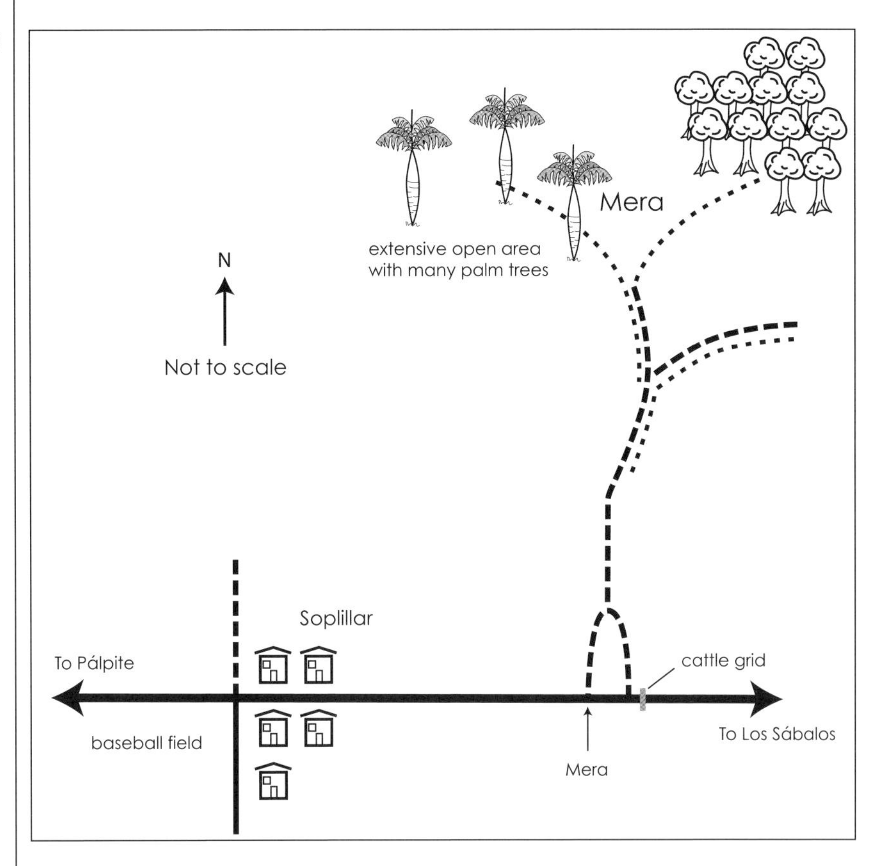


## La Majagua

This spot is about 5km south-east of Soplillar on the left-hand side of the road. You will have to walk into the forest, which can be partially flooded during the rainy season. This is a difficult trail, and it is recommended that you visit it with an experienced guide. Follow the trail for about 100 m to an open area. Most of the herons can be found here, and among the scattered trees and shrubs, look for Grey-headed Quail-Dove, Cuban Parakeet, and several species of warblers and vireos.

## Molina

This is an area of semi-deciduous and second growth forest about 2km south-east of Soplillar. Look for a narrow dirt road on the right-hand side just before you reach an area of open country that has been fenced. Drive the dirt track for about 500m and then walk through the relatively low-canopy woodland. This is an excellent area for Grey-headed Quail-Dove and there are chances for Blue-headed Quail-Dove. Other endemics and near-endemics that can be found in the area include: Gundlach's Hawk (around the clearing after 1km), Cuban Trogon, Cuban Tody, West Indian and Cuban Green Woodpeckers, Crescent-eyed Pewee, Cuban Vireo and Yellow-headed Warbler.

## Los Sábalos

This area of semi-deciduous forest and second growth about 15km south-east of Playa Larga is one of the most profitable birding areas in the entire peninsula. In recent years it has been somewhat 'abandoned' in favour of other areas closer to the main hotels, but if any of the key endemics is eluding you, it will be worth a visit, especially early in the morning when quail-doves are regularly seen on the trails if you are quiet.

***Location***

To reach the area from Playa Larga, drive along the coast road towards Playa Girón and take the second paved road on the left. Drive due north for 5km, passing through the small village of Los Hondones and subsequently past a fenced-off, small military post. Follow the road to the right until you reach an intersection, marked by three houses. Turn right, and after 800m take the first trail on the left that enters the forest (there are palm trees marking the entrance). You can park here. Another means of reaching the area is to take the road that passes through Soplillar, Molina, and La Majagua. You cross a bridge over a canal, turn right and subsequently left, before you eventually reach the previously mentioned house at the intersection of the two roads.

***Strategy***

From the parking area, proceed along the trail, turn right and continue straight ahead until you reach the first left-hand turn. The tall trees here represent one of the best places in Cuba to find Bee Hummingbird, though this entire area is excellent for the species. At this point, you may either continue further or return via the same route. A few metres beyond the left-hand turn, there is a track off to the right. Taking this will bring you back to the main, driveable, road, and by turning right you will reach your starting point. You may also find that it pays to walk even further along the principal track inside the forest, returning via either of the routes mentioned. In addition to Bee

Male Bee Hummingbird

Hummingbird, the area holds many other sought-after species and this is one of the best birding areas within Zapata.

*Birds*

Approximately 90 species have been found in this area. In addition to Bee Hummingbird, the rare Gundlach's Hawk may also be encountered, as well as Broad-winged Hawk, Cuban Parakeet, all four quail-doves (the two rarer species are frequently seen), Great Lizard Cuckoo, Cuban Emerald, Cuban Tody, Cuban Trogon, West Indian and Cuban Green Woodpeckers, Crescent-eyed Pewee, La Sagra's Flycatcher, Red-legged Thrush, Cuban Vireo, Blue-winged, Worm-eating and Yellow-headed Warblers (the first two only in winter or on passage), Cuban Bullfinch and Indigo Bunting (on passage/in winter).

## Los Canales

A complex of canals, created to irrigate rice plantations in the district of Aguada de Pasajeros, and lagoons. Many aquatic species, including West Indian Whistling Duck, can be observed here, as well as along the edges of the road that runs parallel to it. On the opposite side of the road are extensive stands of palm trees, freshwater vegetation, bushes, and sawgrass, offering a good opportunity to see many different bird species.

*Location*

To reach the first canal, follow the previous directions to Los Sábalos from Playa Larga, taking the only unpaved road on the left, or follow the road that passes through Soplillar, Molina, La Majagua, etc. The canal is reached at the end of this road; go over the bridge, turn left and follow the road until you reach open pastures and the first rice fields, approximately 30 minutes from the bridge.

*Strategy*

Explore as much of the wetland complex as time permits. It is worth checking as many areas as you can, especially during passage periods and when water levels are lower, at which times there may be many shorebirds using the lagoons and fields. The open pastures are the best spots for finding migrant sparrows.

*Birds*

Well over 100 species have been recorded at this site, including several rarities. Species to watch for include Least Grebe, Masked and Ruddy Ducks, most herons, Least Bittern, both cormorants, Anhinga, Snail Kite, Gundlach's Hawk (rare), Northern Harrier, Osprey, Merlin, Peregrine Falcon, Sora, Northern Jacana, Wilson's Snipe, Caspian, Royal and Forster's Terns, Barn and Short-eared Owls, Antillean Nighthawk (summer), Belted Kingfisher, Tree Swallow (passage), Cave Swallow (summer), Tawny-shouldered Blackbird, Red-shouldered Blackbird (rare), Eastern Meadowlark, Nutmeg Mannikin (introduced), Indigo and Painted Buntings (passage), Savannah and Grasshopper Sparrows (passage and winter, both are rare). During the rainy season both King Rail and Least Bittern are frequently observed.

## El Cenote

This limestone-based site is about 18km south-east of Playa Larga on the left-hand-side of the road. It is well signed, 'Cueva de los Peces' (Cave of the Fishes) and can be popular with tourists, particularly in the hottest part of the day, as there is a restaurant on site. Most visitors congregate around the large natural pool, which is primarily fresh water but, because an underground channel connects it to the sea, is inhabited by many coral fishes. In the surrounding woodland, a variety of forest species can be found including Blue-headed Quail-Dove, which until November 2001 regularly visited the restaurant in mid-afternoons. More regular fare includes Cuban Bullfinch and Western Spindalis. Migrant warblers and other Nearctic visitors can be common in season, and Bare-legged Owl (until recently a pair occupied a tree directly between the pool and open-air eating area for several years), Cuban Emerald (which has even nested inside the restaurant), Cuban Green Woodpecker, Crescent-eyed Pewee and Cuban Blackbird can also be found here. The refuse tip, which is situated in a small, very degraded area of woodland near a tiny compound directly behind the main restaurant building, and reached by a short, narrow trail, has often held quail-doves in the past, but is increasingly disturbed.

## Bermejas

*Map of Bermejas*

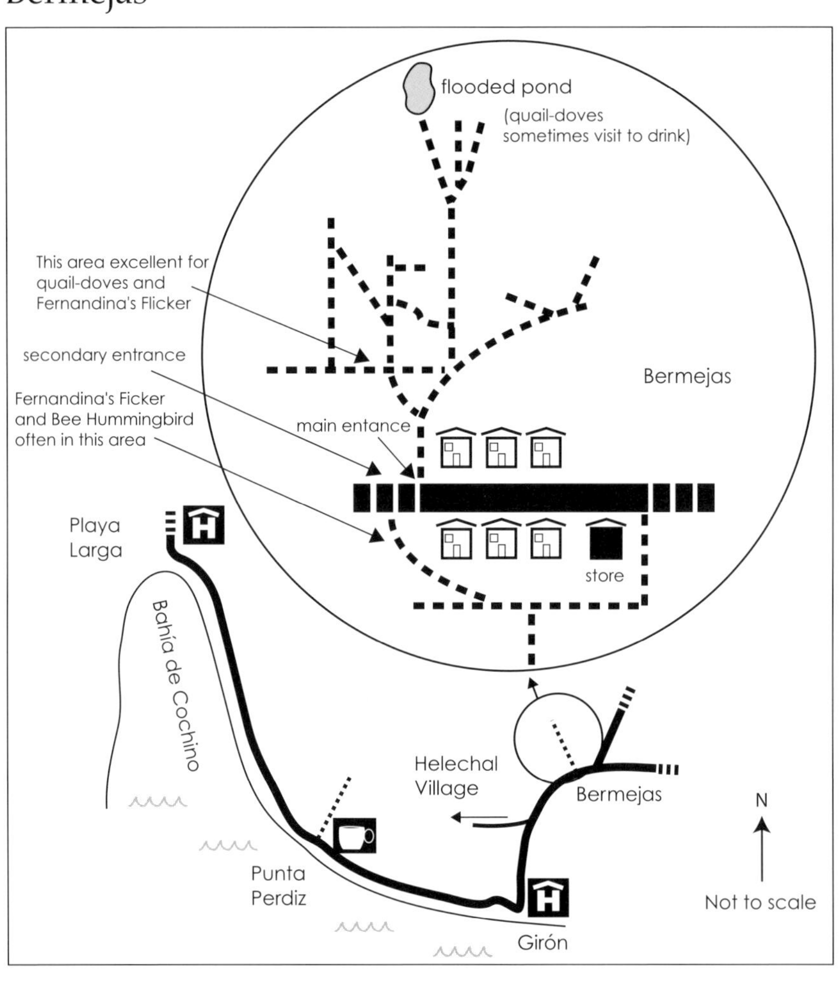

An open area with royal and cabbage palms, second growth, bushes and shrubbery, 12km north of Playa Girón. This is *the* place to look for Fernandina's Flicker, as well as all of the other woodpecker species in Cuba (except Ivory-billed, of course!). Indeed, at the time of writing, late 2009, it can be considered the premier woodland locality in the whole region. Cuba's two endemic owls, Cuban Nightjar, Cuban Parakeet, and all four quail-dove species can also be seen here (the two rarer species are frequently encountered and, since spring 2003, Bermejas has been arguably the best site in the Zapata region for these birds), as well as many Nearctic warblers, and most species of flycatcher found in Cuba. It is also the most reliable spot for Bee Hummingbird in the entire peninsula.

***Location*** Bermejas lies north of Playa Girón on the paved road towards Cienfuegos. After approximately 8km bear right within the village of Helechal and after another 4.1km you reach the settlement of Bermejas. The entrance to the woodland now has a sign above it, and is immediately adjacent to the first house in the village on the left, if driving north. A little further along the road back towards the sea, there is another gated entrance. Park carefully off the road on the wide grassy margins.

***Strategy*** The Bermejas area is now a reserve and is patrolled by a forest guard, to whom you should pay the small entry fee. The guard (at the time of writing, Orlando) knows the whereabouts of a great many of the species unique to the region, and can guide you for an additional small fee, negotiable on site.

On entering the woods via the main entrance, and before reaching the area with cabbage palms, look for Bee Hummingbird among the roadside bushes and flowers. Just a few metres from the road the track reaches a clearing. Cuban Parakeet is often seen around here and over the adjacent village in the evening. Go left at the clearing and upon re-entering forest, take the first track on the left. This area, including the semi-cleared area to the right of the trail is excellent for many hole-nesting species including Rose-throated Parrot (especially in early morning and late afternoon), Bare-legged Owl, Cuban Trogon, and Northern and Fernandina's Flickers, as well as sometimes holding Bee Hummingbird and, regularly, Cuban Tody, Cuban Vireo and Yellow-headed Warbler. Both Grey-headed and Blue-headed Quail-Doves are regular in this area, which can also be accessed from the subsidiary entrance mentioned above. Small waterholes, which usually contain at least some moisture well into the dry season, should be checked early morning and evening for quail-doves, and many other species, coming to drink.

It is possible to continue, beyond the semi-cleared area, along this track for some distance, but the first 500m are best. Re-trace your steps to the first clearing and then take the right-hand trail. This soon enters very productive second growth-forest rich in wintering warblers, Blue-grey Gnatcatcher, and with near-abundant trogons, todies and Great Lizard Cuckoos. After a relatively short distance, another, much larger clearing is reached. This area is superb for Fernandina's Flicker and other woodpeckers, as well as Bare-legged Owl. Cuban Grassquit was formerly regular in this area but has been lost from the region in the last decade. It is worth continuing beyond the clearing, even if you have found all of your target species, the forest in this area can be alive with wintering North American migrants in season, and many of the endemics are easily found here. Blue-headed Quail-Dove is, in some years, regularly seen along this trail. Eventually the trail reaches yet another clearing, which contains a waterlogged area surrounding a clump of trees. This part of the region is especially good for icterids, and may occasionally produce waterbirds.

Back at the main road, and if visiting the area at dusk, take the first right-hand turn when proceeding north through the village, by a small shop, and follow this through the houses to a small clearing in woodland. This and many other areas around Bermejas is a good area for Cuban Nightjar, among them just along the main road (in front of the reserve entrance). Opposite the smaller of the reserve's two entrances is a dirt road (which leads into another patch of woodland between some houses). Cuban Nightjar can be found along here too, and the line of small trees and shrubs immediately by the main road has regularly held Bee Hummingbird and Fernandina's Flicker in recent years.

Another area recommended by visiting birders, but not visited by us, can be reached by continuing north-east beyond Bermejas for 2.4km. At this point take a turn to the right and proceed a further 4km to a small village beyond which the paved road ends after a further 1km. The trail beyond the end of the road proceeds through yet more limestone-based dry woodland with opportunities for quail-doves, including Blue-headed, as well as Gundlach's Hawk and Cuban Parakeet. By taking a short, 250m, trail between the houses, to the west of the village, an area of palms that, at least formerly, supported Fernandina's Flicker can be reached. Anyone having real problems with finding the latter species can also try areas even further north of Bermejas, along the main road towards Cienfuegos. Stop in likely areas with tall palms and suitable nesting holes.

***Birds***

Well over 100 species have been recorded in this area, including the following endemics, near-endemics and other specialties: Sharp-shinned, Gundlach's and Cuban Black Hawks, Northern Caracara, White-crowned Pigeon, Zenaida Dove, Key West, Grey-headed, Ruddy and Blue-headed Quail-Doves, Great Lizard Cuckoo, Cuban Parakeet, Rose-throated Parrot, Bare-legged and Cuban Pygmy Owls, Antillean Nighthawk (summer), Cuban Nightjar, Antillean Palm Swift, Cuban Emerald, Bee Hummingbird, Cuban Trogon, Cuban Tody, West Indian Woodpecker, Yellow-bellied Sapsucker (winter), Cuban Green Woodpecker, Northern and Fernandina's Flickers, Crescent-eyed Pewee, La Sagra's Flycatcher, Loggerhead Kingbird, Cuban Vireo, Black-whiskered Vireo (summer), Red-legged Thrush, 18 species of wood warbler including Yellow-headed Warbler, Western Spindalis, Cuban Bullfinch, Cuban and Red-shouldered Blackbirds, and Cuban Oriole.

## La Salina

*Map of Salina*

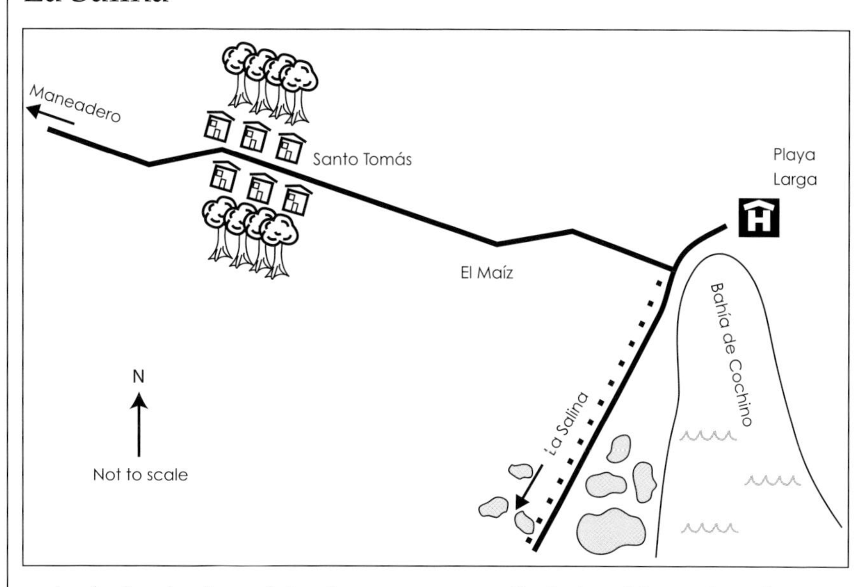


At the beginning of the dry season, usually in late November, this can be one of the most impressive areas for birding in Zapata. At this period the wetlands are drying out, leaving only scattered puddles and stagnant, shallow waters. These are ideal feeding conditions for many waterbirds, which come to feast on the fishes, crustaceans, and other invertebrates. If you are lucky to be in the area at this time, you will have the good fortune to visit one of the richest birdwatching areas in the West Indies.

***Location***

Reaching La Salina is simple. The abandoned saltpans are about 26km south of Buena Ventura, a small village just 3km west of Playa Larga. From the entrance to the hotel at Playa Larga turn left, as if driving to Havana, and then left again into the village at the obvious bend in the road after a just a couple hundred metres. Continue through Playa Larga along the dirt road to Buena Ventura. Because La Salina is a national sanctuary, you must first obtain permission to visit from the local authorities, at the Agriculture Ministry office in Playa Larga, and then contract a local guide (this can also be arranged by the park office), who will open the sanctuary gate. Take the left turn immediately beyond the gate (straight ahead is Santo Tomás).

***Strategy***

The area around the Agriculture Ministry office and nearby police station is an extremely regular locality for Cuban Crow. For several kilometres beyond Buena Ventura, there is open forest on limestone-rich soil. Many of the commoner resident birds can be seen here, but look especially for Cuban Black Hawk, which is very tame here. Also possible are Gundlach's, Red-tailed or Broad-winged Hawks. At night along this road is a reliable area for Bare-legged Owl (your local guide may know a nest or roost site close to the road) and Cuban Nightjar.

Soon the mangrove gives way to more open, savanna-like terrain. Look for Sandhill Crane feeding here, especially in the early morning (note, however, that there appear to have been very few, if any, recent sightings of the species). Once beyond the savanna, you enter a wetland area very typical of the Zapata region. The greatest diversity of waterbirds occurs here. Many species that in other areas are occasionally seen singly, or not at all, can be abundant including American Flamingo, Roseate Spoonbill, both cormorants, many herons including Reddish Egret, Osprey, Caspian Tern, White Ibis and Wood Stork. With few exceptions, all of the species of sandpipers and plovers on Cuba can be abundant. Look for Peregrine Falcon, Clapper Rail (common but usually requires playback to lure into view) and Yellow Warbler as well. The road ends at an abandoned saltpan, where Wilson's Plover breeds. Occasionally, American Crocodile may also be seen in this area.

*Birds*

More than 120 species have been recorded at this site including a number of commoner endemics, but the main interest is waterbirds and the chance of scarce migrants. Some of the species of most interest include: American White Pelican (a rare winter visitor), American Flamingo, Anhinga, Yellow-crowned Night Heron, White Ibis, Roseate Spoonbill, Wood Stork, Osprey, Gundlach's, Cuban Black, Broad-winged and Red-tailed Hawks, Clapper Rail, Limpkin, Sandhill Crane, Wilson's Plover, Gull-billed, Caspian and Royal Terns, White-crowned Pigeon, White-winged and Zenaida Doves, Rose-throated Parrot, Mangrove and Great Lizard Cuckoos, Bare-legged and Cuban Pygmy Owls, Antillean Nighthawk (summer), Cuban Nightjar, Cuban Emerald, Cuban Trogon, Cuban Tody, Belted Kingfisher, West Indian Woodpecker, Yellow-bellied Sapsucker (winter), Cuban Green Woodpecker, Crescent-eyed Pewee, La Sagra's Flycatcher, Grey Kingbird (summer), Loggerhead Kingbird, Cuban Vireo, Black-whiskered Vireo (summer), Cuban Martin (summer), Red-legged Thrush, Yellow-headed Warbler, Western Spindalis, Cuban Bullfinch, Cuban Blackbird and Greater Antillean Grackle.

## Santo Tomás

This area is the site where the Spanish naturalist, Fermín Cervera, discovered three of Cuba's most famous endemics – Zapata Wren, Zapata Sparrow and Zapata Rail – in 1926. Following the official description of these species the area became famous. Surrounding the marsh is wet forest with such sought-after birds as Gundlach's Hawk, Grey-headed Quail-Dove and Bee Hummingbird. Visiting Santo Tomás requires special permission from the local forestry authorities, the Agriculture Ministry office just north of Playa Larga (the cost is currently is 10 CUC/US$11), and the services of a guide in this area are also compulsory. None of the species found here is unique to the site, and all can be found with arguably greater ease at other sites but this is nonetheless an interesting area to visit, especially because of the historical context.

*Location*

Prior to 1960, it was practically impossible to reach Santo Tomás by land, as it is 36km from Playa Larga and lies almost in the heart of the swamp. To reach the area follow the same initial directions as if going to La Salina, but at the control gate, instead of taking the left-hand road, keep going straight until you reach the village of Santo Tomás. At the local store, known as a 'bodega', take the narrow dirt road to the right (passing an outdoor oven used to make the charcoal), until you reach its end at a canal.

***Strategy*** Along the road are semi-deciduous and evergreen woods, but the birds inhabiting them are also found in other forests. We advise against the temptation to stop en route. It is necessary to reach Santo Tomás during the first hour of daylight, before the wind has picked up in the exposed sawgrass wetlands. Cutting through the marshland is a canal, La Cocodrila, formerly used as a route to extract timber from the woods. The best way to reach the marshes is to take a small boat along the canal, listening for the distant song of Zapata Wren, or watching for any Zapata Sparrows. The rail is also found in this area, but is practically impossible to see due to its secretive habits (the same is true at all the other sites where it has been very occasionally seen). The other way to enter the sawgrass wetland is to walk from the village. During the dry season this route is feasible; during the wet season, however, the water level is high, up to 1.5m deep in places, and there are many sink holes, difficult to detect when the water level is high. You are almost sure to end up entirely soaked! Afterwards, look for Bee Hummingbird among the trees and gardens of the village houses. Another target species here is Sandhill Crane, usually best located just to the north of the entrance of the reserve, at the point with the road barrier and control guard. Just walk about 70m along the narrow trail north of the road to watch for the cranes, which are usually present only early in the morning (and, as elsewhere in the Zapata region, their current status is unknown).

***Birds*** More than 100 species have been recorded in the environs of the village, including almost 20 endemics. Almost all of the rails reported in Cuba occur in these wetlands, and some can usually be heard calling, especially King and Spotted Rails. This is also a good place to find Limpkin, Glossy Ibis, Wilson's Snipe, and both night herons. Blue-winged Teal, gallinules, Masked and Wood Ducks, especially the latter, can sometimes be found along the canal, as well as herons, cormorants, and a few species of warblers (e.g. waterthrushes in winter). Common Yellowthroat, Crescent-eyed Pewee, and kingbirds are rather common, and occasionally Bee Hummingbird can be found at the entrance of the canal, along with Cuban Tody, Yellow-headed Warbler, Cuban Vireo and Cuban Bullfinch. The area is also excellent for Gundlach's Hawk and Grey-headed Quail-Dove. Your guide should have up-to-date information on the whereabouts of these species.

## Guamá

This is a tourist resort, consisting of a group of cabins, restaurants, bars and bridges on the waters of Treasure Lake (Laguna del Tesoro).

There is also an Indian village with stone statues depicting the former inhabitants of the region, the Taíno Indians.

***Location*** Treasure Lake comprises two parts: La Boca and Guamá. The first is by the main road, 12km north of Playa Larga and forms the entrance to the lake. There are souvenir shops, a bar and a restaurant.

***Strategy*** La Boca's gardens attract many species of birds, including Rose-throated Parrot, Cuban and Tawny-shouldered Blackbirds, Greater Antillean Grackle, Cuban Oriole, and very infrequently Red-shouldered Blackbird. Cuban Crow is regular in the taller trees around the tourist buildings. However, some of the area's attractiveness to birds has been lost in recent years, as many of the taller trees have been destroyed by hurricanes that ripped through the area in the autumns of 2001 and 2002. In the aquatic vegetation, look for Least Bittern, Northern Jacana and gallinules. Behind the restaurant is a large crocodile farm, where about 12,000 crocodiles (most of them the endemic *Crocodylus rhombifer*) are raised. Cuban Pygmy Owl, as well as warblers, woodpeckers (sometimes including Fernandina's Flicker), and kingbirds can be seen around the farm.

After exploring La Boca and its surroundings, you could take a boat trip along the canal and across the lake to the resort of Guamá, which takes about 35 minutes. Scan the exotic *Casuarina* trees that fringe the banks for Osprey, Gundlach's Hawk (best searched for from the tower above the restaurant at Guamá), Antillean Palm Swift and Crescent-eyed Pewee.

At Guamá, search the trees of the resort carefully for Stygian Owl. Bee Hummingbird rarely visits the bottlebrush flowers (and at La Boca). Along the interior canals of Guamá several waterbirds can be found, including Wood Duck, as well as Snail Kite, which can usually be seen perched at the end of the canal or patrolling the lake in search of snails. Along the lake's edge, look for Neotropic Cormorant and Anhinga.

***Birds*** The following species of potential interest are possible in this area: Least and Pied-billed Grebes, Yellow-crowned Night Heron, West Indian Whistling Duck, Wood Duck, Osprey, Snail Kite, Northern Harrier, Gundlach's and Cuban Black Hawks, Purple Gallinule, Limpkin, Northern Jacana, White-crowned Pigeon, Rose-throated Parrot, Great Lizard Cuckoo, Barn, Cuban Pygmy and Stygian Owls (rare), Antillean Nighthawk (summer), Cuban Emerald, Bee Hummingbird (rare), West Indian and Cuban Green Woodpeckers, Northern and Fernandina's Flickers, Crescent-eyed Pewee, La Sagra's Flycatcher, Cuban Crow, Cuban Martin (summer), Red-legged Thrush, many wood warblers, including Yellow-throated and Prairie (check the bottlebrush trees), Cuban Bullfinch, Yellow-faced Grassquit, Tawny-shouldered and Cuban Blackbirds, Greater Antillean Grackle and Cuban Oriole.

## Pálpite

The semi-deciduous and evergreen woodland surrounding this

settlement is excellent for a broad selection of species, including many endemics.

***Location*** Approximately 5km south of Guamá is the village of Pálpite. Upon entering the village from the north take the first track on the right-hand side of the road, and park, after about 100m, immediately beyond the last house on the right. Continue on foot into the forest. After just a few metres there is a three-way fork in the trail. By going either straight ahead (from where a number of side trails emanate, all of which are good for birding, although some are too overgrown to readily find quail-doves) or left, you enter excellent birding habitat, seasonally inundated limestone woodland. In our experience the left-hand track is consistently the most productive, while a trail off to the right, about 1km along this track, was also worthy of exploration prior to Hurricane Michelle, in early November 2001, but many of the trees in this area were killed during the storm. It is possible to walk at least 5km along the left-hand trail, but in practice it is probably not worth going more than 2km along here.

***Map of Pálpite***

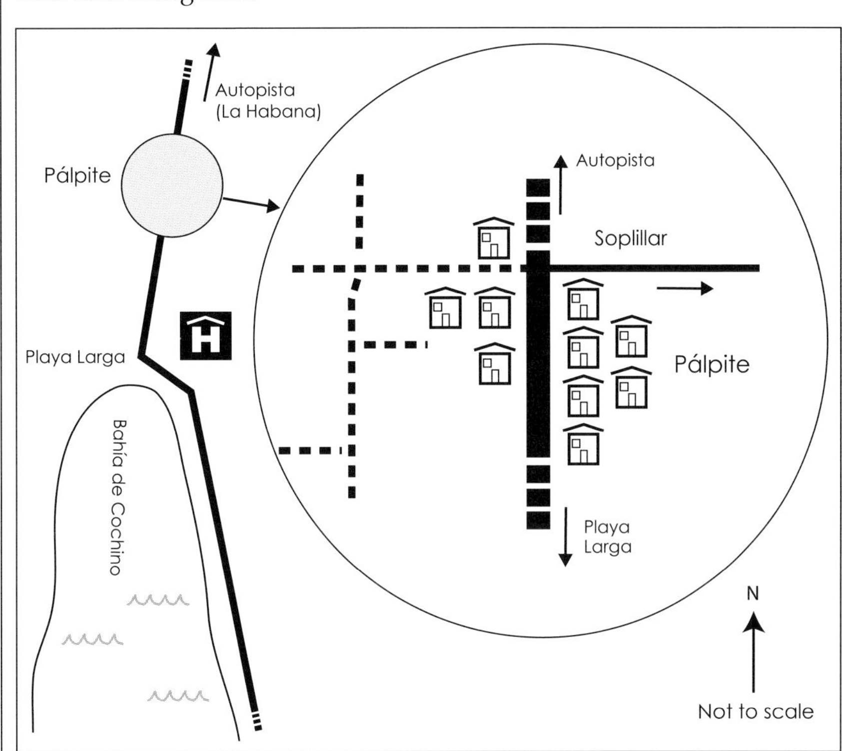


***Strategy*** This area can be superb for quail-doves, with all four species present. Most numerous are Ruddy and Key West Quail-Doves, but Grey-headed Quail-dove is also encountered relatively frequently, while Blue-headed Quail-Dove is not uncommon, although this is not the best site for the species. Virtually any time of day can be productive, and with a significant slice of good fortune it would be possible to record all four during a day. In practice, a number of visits, at different times of day, will be necessary for an observer to see three or four species. Zenaida

Dove is also common in the area. As mentioned, a wide range of endemics and near-endemics occur here, with White-crowned Pigeon, Rose-throated Parrot, Cuban Trogon, Cuban Tody, Crescent-eyed Pewee, La Sagra's Flycatcher, Red-legged Thrush, Black-whiskered Vireo (March–October), Cuban Vireo and Yellow-headed Warbler, and Cuban Tody among the most frequent. Bee Hummingbird is regular along the left-hand trail, in tall, dead growth between the main fork and the first track off to the right, and in a small clearing (the product of Hurricane Michelle) 500m beyond the latter, while Cuban Emerald is also, unsurprisingly, common. In winter, the area is a haven for many Nearctic migrants, including Yellow-bellied Sapsucker, Grey Catbird and almost innumerable North American warblers.

Cuban Crow is frequent around the village, as is Antillean Nighthawk, in season, and all of Cuba's icterids, except Red-shouldered Blackbird. The nighthawk should also be searched for over open ground, on both sides of the main road, just north of Pálpite, and along the dirt road east of the village. There is a small roost of Rose-throated Parrots in the village, and, in season, Cuban Martin is regularly recorded over the settlement early morning and evening. Cuban Parakeet is sometimes regular in summer. Cuban Nightjar and Cuban Pygmy Owl are reasonably regular along the first dirt road running east of the main road in the village, and may also be found elsewhere in the area, while Stygian Owl has been recorded in the village, and this is one of the few known localities for Northern Potoo in Cuba (although there have only been a couple of records to date). Bare-legged and Stygian Owls can be found in the evening around the CITMA station just south of the village on the right-hand (west) side of the road to Playa Larga.

Any trail in the area is worthy of exploration; in April 2000, GMK recorded both rarer quail-doves along a short trail just south and east of the village in the same evening. There are also several small pools in the area, the largest being a short distance along the main dirt road east from the village, towards Soplillar, which occasionally produce waterfowl and shorebirds, although they are to some extent seasonal. However, observers should concentrate on the trail west of the village described above.

***Birds***

Almost 120 species, including 17 endemics, have been recorded in the area including the following species of particular interest: Sharp-shinned and Gundlach's Hawk (both rare), White-crowned Pigeon, Zenaida Dove, Key West, Grey-headed, Ruddy and Blue-headed Quail-Doves, Cuban Parakeet, Rose-throated Parrot, Great Lizard Cuckoo, Bare-legged, Cuban Pygmy and Stygian Owls, Common Nighthawk (passage), Antillean Nighthawk (summer), Northern Potoo (rare), Cuban Nightjar, Cuban Emerald, Bee Hummingbird, Cuban Trogon, Cuban Tody, West Indian Woodpecker, Yellow-bellied Sapsucker (winter), Cuban Green Woodpecker, Crescent-eyed Pewee, La Sagra's Flycatcher, Loggerhead Kingbird, Cuban and Black-whiskered Vireos (summer), Cuban Crow, Cuban Martin (summer), Red-legged Thrush, Yellow-headed Warbler, Red-legged Honeycreeper (rare), Western

Spindalis, Cuban Bullfinch, Cuban Blackbird, Greater Antillean Grackle and Cuban Oriole.

## La Turba

This area of swampy forest is the most accessible site on the peninsula for three keynote endemics, Zapata Wren, Zapata Sparrow and Red-shouldered Blackbird. As such it is an essential area for the keen endemic hunter, and should be targeted for an early-morning visit, before any wind gets up (after which the wren, in particular, will fall silent and be nigh-on impossible to find).

***Location*** North of Guamá is an often very productive area of marshland known as Turba. If coming from the south, about 5km beyond Guamá you will note a dirt road off to the left (west). Immediately beyond the dirt road is a compound, with a sign 'Turba, Zapata' and then a police checkpoint with a 'Welcome to the Zapata Peninsula' sign, which is never manned. If you reach these you need to turn round, although immediately beyond the checkpoint, on the right-hand (east) side of the main road there is a line of royal palms, which has hosted Fernandina's Flicker, whilst the adjacent wet area, if planted with rice, may produce Red-shouldered Blackbird. If dry, or no rice has been planted, the blackbird will be absent, but you may be able to continue as far as a track through reasonable reasonably productive woodland, and the scrubby areas appear potentially productive for sparrows in winter. Other areas, a little further north of the checkpoint, and also to the east of the road, may also be worth checking for the woodpecker.

***Strategy*** The road to Turba at first passes through second-growth woodland. In winter (October to late March) it is worth checking for Chuck-will's-widow, and for Cuban Nightjar at any season. After about 0.75km there is a sharp turn-off to the left (this road occasionally produces Fernandina's Flicker). Continue straight ahead, and after 2.5km the road bends to the left; for the next few kilometres the road, which is tree-lined, passes through marsh, with a deep channel on the right. This area regularly produces herons, Northern Harrier (winter only), and Belted Kingfisher, while Gundlach's Hawk is occasionally recorded. Check for Zapata Wren along here and along the trail off to the left halfway between the sharp bend to the left and the fish farm. At the end (8km from the main road) there is a gate and a small fish farm. This area supports Limpkin, Purple Gallinule, both waterthrushes, Zapata Wren (which has apparently only recently colonised the area; its obvious song cannot previously have gone unnoticed by the many birders familiar with the species' vocalisations that have visited the area), and Zapata Sparrow, whilst both Swainson's Warbler and Lincoln's Sparrow have been recorded in winter. Turba also has the capacity to produce surprises; in recent years both American Goldfinch and American Black Duck have been reported (full details of the latter, the only Cuban record, are still unpublished).

Continue, on foot, beyond the fish farm. There are deep channels on both sides of the broad track; this area is superb for rallids and the use of tapes will be of real benefit. Yellow-breasted Crake, and Spotted and King Rails are all recorded with some frequency. Nonetheless a considerable degree of fortune will be required to see any of these, with the exception of King Rail. Sora is also frequently heard in winter, and American Bittern can be expected to occur at the same season. In addition to species previously mentioned, the area just beyond the fish farm also harbours Anhinga, Least Bittern, many ducks (Blue-winged Teal is probably most frequent), passage shorebirds in season if the water level is sufficiently low, hirundines (particularly during passage periods), and Red-shouldered Blackbird.

It should be noted that access to this area is occasionally suspended. Local guides can advise on the current situation, but should such a situation exist it may be worth visiting the crocodile breeding area, west of the main road to Playa Larga, south of the entrance track to Turba. Access to this area can usually be obtained at the gate. By walking the dykes between the ponds you will have chances of Fernandina's Flicker and Red-shouldered Blackbird, although the number of waterbirds in this area is considerably less than at Turba. Masked Duck may also be found here.

***Birds***

Possibilities at this site include the following: Anhinga, Least Bittern, Wood Stork, West Indian Whistling Duck (rare), Wood Duck, Blue-winged Teal, Snail Kite, Northern Harrier (winter), Gundlach's Hawk (uncommon), King Rail, Sora, Yellow-breasted Crake (uncommon), Spotted Rail (common by voice), Purple Gallinule, Limpkin, Northern Jacana, Zenaida Dove, Great Lizard Cuckoo, Cuban Pygmy Owl, Antillean Nighthawk (summer), Chuck-will's-widow (winter), Cuban Nightjar, Fernandina's Flicker (uncommon), Crescent-eyed Pewee, La Sagra's Flycatcher, Zapata Wren, Red-legged Thrush, Northern and Louisiana Waterthrushes, Yellow-headed Warbler, Indigo Bunting (winter), Zapata Sparrow (rare), Red-shouldered, Tawny-shouldered and Cuban Blackbirds, and Greater Antillean Grackle.

## Peralta

This site has the distinction of being one of the few areas where the near-mythical Zapata Rail has been seen in the last decade. This extraordinarily difficult bird aside, the area has a good range of other endemics, but is far from any of the region's hotels.

***Location***

Beside the highway between Havana and Santa Clara, Peralta is located at km122, or from Zapata towards Havana it is about 20km west of the entrance road to Zapata. Being within the national park, a guide is necessary to visit this area. At the start of the trail there is a reddish iron gate (which indicates the entrance).

***Strategy***

The trail is about 2.5km long, and a broad selection of forest species, including Fernandina's Flicker, Cuban Trogon, Yellow-headed Warbler and Cuban Pygmy Owl, are present. Most of these are common, the trogon even occurs in low trees within the more swampy areas. The best area for the flicker is around the Royal Palms at the entrance of the trail, but it is also frequently recorded along the first 500m and around the large, obvious clearing after 1km. The area is also good for Cuban and Red-shouldered Blackbirds. Migrant warblers, particularly Common Yellowthroat and both waterthrushes are common in the dense, frequently waterlogged, track-side vegetation. Also, by walking this trail early in the morning (which we recommend, as it is essential to visit this area very early in order to find the key species of the region), it is possible to find Ruddy and Grey-headed Quail-Doves on the path. But, the main targets of this area are the three rare endemics of the swamp: Zapata Wren, Zapata Sparrow, and Zapata Rail. After about 2km you will notice the habitat transition from forest to swamp vegetation. These three species are present only in the swamp, with the best access to the marsh being on the right-hand side of the trail. Just walking into the swamp for about 70m it is possible to hear and see the wren, although by using a tape it may be possible to bring one to the trail edge. The sparrow can be detected in vegetation along the trail, but also in the swamp. It is very secretive and its slightly buzzing song is low-pitched and easily overlooked, but, once learnt, provides an easy clue as to its presence. To find the rail is exceptionally hard work, but there is no doubt that, in the early part of the current decade, this area was the best place to try to find it. The marshland is also excellent for Red-shouldered Blackbird as well. Spotted and King Rails are also present.

***Birds*** Possibilities at this site include the following: Northern Harrier, Gundlach's Hawk (rare), King Rail, Sora, Zapata (extremely rare; no sightings in the last few years), and Spotted Rails (common by voice), Limpkin, White-crowned Pigeon, Zenaida Dove, Grey-headed and Ruddy Quail-Doves, Cuban Parakeet (uncommon), Rose-throated Parrot, Great Lizard Cuckoo, Cuban Pygmy Owl, Cuban Nightjar, Cuban Emerald, Bee Hummingbird (rare), Cuban Trogon, Cuban Tody, West Indian and Cuban Green Woodpeckers, Fernandina's Flicker, Crescent-eyed Pewee, La Sagra's Flycatcher, Grey and Loggerhead Kingbirds, Cuban and Black-whiskered Vireos (summer), Zapata Wren, Red-legged Thrush, Northern and Louisiana Waterthrushes, Yellow-headed Warbler, Cuban Bullfinch, Zapata Sparrow, Red-shouldered and Tawny-shouldered Blackbirds, Eastern Meadowlark (fields opposite the entrance to Peralta, on the other side of the autopista), Cuban Blackbird, Greater Antillean Grackle and Cuban Oriole.

## Hato de Jicarita

This is a wonderful marshland, where it is possible, using a boat, to reach the heart of the swamp. As with other localities situated within the national park, permission and the services of a guide, both of which can be obtained from the Agriculture Ministry office in Playa Larga village, are essential.

*Location*

The locality lies south of the highway between Havana and Cienfuegos, at km101. Turn south (right if coming from Havana) onto road intersecting the highway near the kilometre post. (There is also a small service area nearby on the north side of the road.) Drive along the paved road for 1km, before turning left onto a dirt road, and drive for 6km until you reach its end at a forest guard station. It is possible to rent small boats along the Hatiguanico River here.

*Strategy*

Just 200m before the end of the road, both Zapata Wren, Zapata Sparrow and Red-shouldered Blackbird can be found in the sawgrass beside the track. Red-shouldered Blackbird also occurs around the guard station. A boat trip in this area is always worthwhile, offering close views of Snail Kite, many egrets, Rose-throated Parrot, Gundlach's Hawk, Northern Jacana, Limpkin, and others, although the main targets will once again be the wren and sparrow. The marshland habitat south of the river after 300m is one area to find these species, although the area close to the landing stage about 3km along the river is best for the wren, and this area also regularly produces Gundlach's Hawk.

*Birds*

Possibilities at this site include the following: Least and Pied-billed Grebes, Anhinga, Osprey, Snail Kite, Sharp-shinned and Gundlach's Hawks, King Rail, Sora, Zapata and Spotted Rails (both extremely difficult), Purple Gallinule, Limpkin, White-crowned Pigeon, Rose-throated Parrot, Great Lizard Cuckoo, Cuban Pygmy Owl, Cuban Emerald, Cuban Trogon, Cuban Tody, Belted Kingfisher, Cuban Green Woodpecker, Crescent-eyed Pewee, La Sagra's Flycatcher, Grey and Loggerhead Kingbirds, Zapata Wren, Red-legged Thrush, Zapata Sparrow, Red-shouldered and Tawny-shouldered Blackbirds, Eastern Meadowlark, Greater Antillean Grackle and Cuban Oriole.

## Punta Perdiz

For those staying in Playa Girón, this site and Bermejas make good early-morning destinations. The facilities at Playa Girón are very similar to those at Playa Larga, though in our opinion better. However, many of the best birding sites are closer to the latter resort. Those staying at Playa Girón should search for the resident pair of Stygian Owls, which are best looked for on dark nights around either of the two swimming pools in the complex, or in trees close to the entrance barrier and disco! In summer, Antillean Nighthawks can frequently be seen at dusk around the resort, sometimes coming in off the sea earlier in the spring, and at any time it is worth checking the little harbour for Wilson's Plover.

*Location*

The Punta Perdiz trail is located beside the main road from Playa Larga to Playa Girón. Approximately 7km west of Girón, there is a restaurant on the seaward side of the road, incongruously shaped like a boat, and 80m further towards Playa Larga, on the right-hand side of the road, there is wide trail entering the woods. If coming from the west (Playa Larga) there is a large sign, announcing Punta Perdiz, immediately before and on the same side of the road as the trail.

***Strategy***

This can be a good place to find quail-doves, principally Key West, but Ruddy and Blue-headed Quail-Doves are also seen. Several of the common endemics, e.g. Cuban Trogon, Cuban Pygmy Owl, Cuban Tody, and Yellow-headed Warbler are also easily found in this area, and Nearctic migrants are common throughout in winter. Large roosts of White-crowned Pigeons have been recorded near the road in early summer.

***Birds***

Possibilities at this site include the following: Gundlach's Hawk (rare), White-crowned Pigeon, Zenaida Dove, Key West, Grey-headed, Ruddy and Blue-headed Quail-Doves, Cuban Parakeet, Rose-throated Parrot, Great Lizard Cuckoo, Bare-legged, Cuban Pygmy and Stygian Owls, Chuck-will's-widow (winter), Cuban Emerald, Bee Hummingbird (rare), Cuban Trogon, Cuban Tody, West Indian Woodpecker, Yellow-bellied Sapsucker (winter), Cuban Green Woodpecker, Northern and Fernandina's Flicker (both rare), Crescent-eyed Pewee, La Sagra's Flycatcher, Loggerhead Kingbird, Cuban Vireo, Black-whiskered Vireo (summer), Red-legged Thrush, Yellow-headed Warbler, Red-legged Honeycreeper (rare), Western Spindalis, Cuban Bullfinch and Cuban Oriole.

# The West

## Guanahacabibes

Lying 380km west of Havana, with an area of 210km$^2$, the north coast of the Guanahacabibes Peninsula, which forms the westernmost tip of Cuba, consists of muddy shores with several lagoons. The peninsula, which is a designated national park (in 2001) and Natural Biosphere Reserve (established in 1987), is one of the most naturally pristine areas of the country. The area has several habitats, with semi-deciduous woodland and coastal thicket dominant, and is of greatest ornithological importance for the large numbers of Nearctic migrants in winter and during passage. Access to the national park is controlled, with a barrier at La Bajada, near the base of the peninsula. Permits to proceed beyond the checkpoint can be obtained (for a small fee) from the national park office on the right-hand side just before it, where you can also arrange a guide to accompany you. There is a second checkpoint at Faro Roncali.

***Location***

Follow the autopista to its terminus in La Fe and from there follow the signs to Villa María La Gorda. Eventually the national park office at La Bajada is reached; by following the road around to the left you will reach the resort; straight ahead, beyond the barrier, is the road out across the peninsula.

***Accommodation***

The closest facility is Villa María La Gorda, in the south-east of the peninsula. Pre-booking is advisable as this relatively small resort is very popular with scuba divers. It is possible to arrange package tours from Havana, at a price. There is at least one *casa particular* in the area, but you will have to ask around to find it.

*Strategy*

Autumn migration is spectacular (particularly 10 October–10 November), with the constant possibility of a new bird species for Cuba (and even the West Indies). The most recent discovery was in autumn 2009, when one of us (Kirkconnell) found the first Swainson's Hawk in Cuba, which was only the second record for the Antilles (and first to be documented with photographs). Spring migration is generally much quieter. Overall diversity is very high: over 190 species have been reported. Of special interest are Blue-headed Quail-Dove, Rose-throated Parrot, Cuban Pygmy and Bare-legged Owls, Bee Hummingbird, Cuban Trogon and Cuban Green Woodpecker. In winter, the following migrants can be present in large numbers: Northern Parula, Black-throated Blue and Worm-eating Warblers, Blue Grosbeak and Savannah Sparrow. Cerulean and Hooded Warblers may also be present. The entire forested area is good for birding, but exceptional for both residents and migrants are those woodlands closest to the coastline accessed via the main road along the south of the peninsula (any trail could be productive), and the lighthouse at Faro Roncali, as well as the scrub near the tip of the peninsula and the 3km of coast at Playa Las Tumbas (Cabo de San Antonio), which marks the end of the road. Note the road is rather rough and should be driven with care; four-wheel drive would be advantageous. There is a small network of trails immediately behind the national park office, where Bare-legged Owl has been found breeding in a cave, and at the far end of the María La Gorda resort. Bee Hummingbird is regularly found at a locality known as El Berraco, near Punta Gorda, and close to the resort. It is possible to hire a national park guide; some of the staff are reasonably knowledge about birds.

*Birds*

The many possibilities include the following: Yellow-crowned Night Heron, West Indian Whistling Duck, Gundlach's Hawk, Clapper and King Rails, Limpkin, Wilson's Plover, Scaly-naped, White-crowned and Plain Pigeons (occurs around Villa María La Gorda), Ruddy and Blue-headed Quail-Doves, Rose-throated Parrot, Great Lizard Cuckoo, Bare-legged, Cuban Pygmy and Stygian Owls, Antillean Nighthawk (summer), Chuck-will's-widow (winter), Cuban Nightjar, Antillean Palm Swift, Cuban Emerald, Bee Hummingbird (can be common around Villa María La Gorda), Cuban Trogon, Cuban Tody, West Indian and Cuban Green Woodpeckers, Crescent-eyed Pewee, Loggerhead Kingbird, Giant Kingbird (rare), Cuban and Black-whiskered Vireos (summer), Cuban Crow (common on the approach road), Cuban Martin, Red-legged Thrush, 30 species of wood warblers, Red-legged Honeycreeper, Western Spindalis, Cuban Bullfinch, Cuban Grassquit (current status unknown), Tawny-shouldered Blackbird, Eastern Meadowlark, Cuban Blackbird, Greater Antillean Grackle and Cuban Oriole.

## La Güira

This national park, in the Sierra de los Organos, lies 150km west of Havana. The closest town is San Diego de los Baños. The main habitats are semi-deciduous woodland, gallery forest, tropical karstic forest and pine forest, and the principal species of interest are Cuban Solitaire and Olive-capped Warbler.

***Map of La Güira***

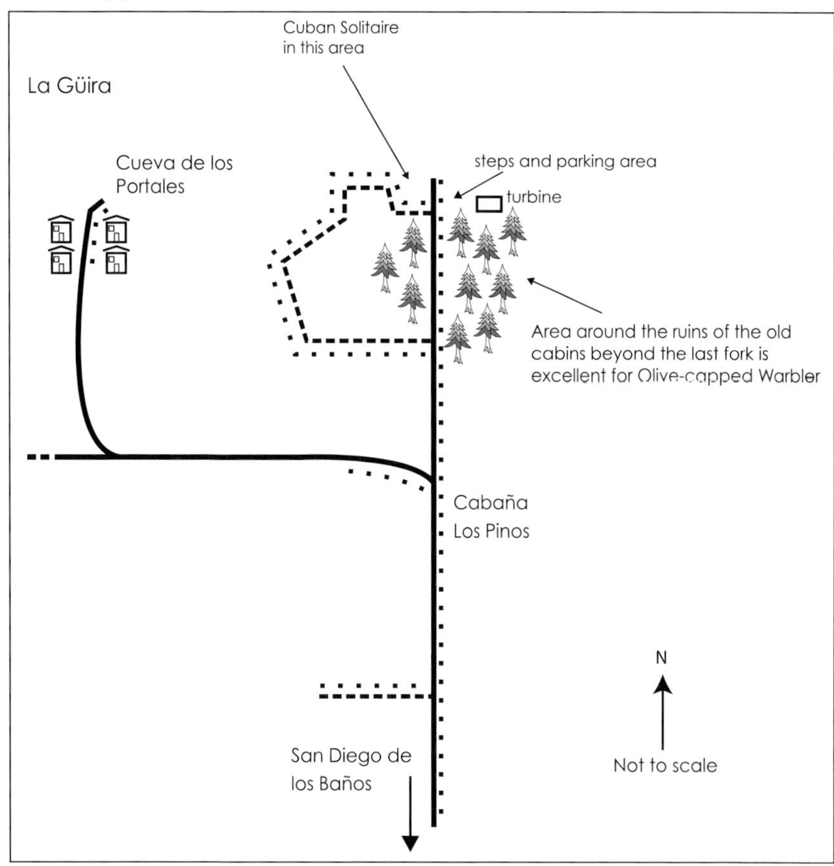


***Location***

San Diego is just 12km from La Güira. Leave the autopista northwards to San Diego and on reaching the crossroads in the centre of the town (to the left is the Hotel Mirador) turn left, go past the main square, and take the next right turn, which continues to the national park. The rather grandiose entrance is on the right. Continue uphill, following the main road all the time; on reaching the pine zone and prominent fork, go right and continue to the end of the road past some ruined chalets, where you should park.

***Accommodation***

You can day-trip La Güira either from Soroa, less than an hour away, or from Havana, which is less than two hours distant, or stay at the Hotel Mirador or any of the *casas particulares* in San Diego de los Baños.

***Strategy***

The best birdwatching trails at La Güira are close to each other. The first trail runs downhill, on the left-hand side of the parking area and to the right of the old turbine (see map). Cuban Solitaire is easily detected by its unmistakable song. Watch out for Cuban Trogon, Ruddy Quail-Dove and Scaly-naped Pigeon. The second path leaves from beside a cave, downhill from the turbine. From here, there are two paths with several intersections. Explore these for Blue-headed and Grey-headed Quail-Doves, Cuban Solitaire, and further downhill is an intersection with the road where you can easily find Western Spindalis, Olive-capped Warbler, Red-legged Honeycreeper, and perhaps even a soaring Gundlach's Hawk. Watch for Scaly-naped Pigeon in flight. There are also smaller paths at both sides of the road.

When leaving the cabins downhill, search the low vegetation just beyond the lawn: if water is dripping on it, up to 16 species can be found foraging here. On the third trail, you may find many warblers, including Olive-capped, and Red-legged Honeycreeper.

Leaving the pine forest, turn right to the Cuevas de los Portales, where Cave Swallow breeds in March to August. There is a small entrance fee payable here, as this is a historic site, famous because Che Guevara spent the period of the so-called 'Cuban missile crisis' here. This is also a good site for Cuban Solitaire, Cuban Tody and Scaly-naped Pigeon. Nearby, it is possible to find Cuban Grassquit. At the intersection with the main road, where you turn left to return to La Güira, turn right and walk the road; the grassquit occurs in the pine forest here. However, access to this area is sometimes curtailed by the Cuban military, but it may is still be worth checking any areas of similar habitat in the region for this increasingly localised species in western Cuba.

Returning towards San Diego, about 1km downhill from the cabins there is a small bridge over a stream; about 50m before this is a trail on the right which should be searched for Ruddy and Blue-headed Quail-Doves, Louisiana Waterthrush, Worm-eating Warbler and the secretive Swainson's Warbler. The best place in the area to find Blue-headed and Key West Quail-Doves is 200m beyond the bridge, and on the right, on a bend in the road, is a trail.

Giant Kingbird has recently been found near the Hotel Mirador. Just 40m downhill is a turning on the right to the thermal baths. The kingbird has been found beside the river here. By walking diagonally across a park by the entrance to the baths you reach an old wooden bridge. On the other side, turn right and walk to the forest. From here it is possible to follow a loop trail (about 1 hour): at the first fork take the left-hand path, which takes you back to a small farm, from where a dirt road returns to the bridge. Birds here include Key West, Ruddy and Blue-headed Quail-Doves, Cuban Pygmy Owl, Giant Kingbird, Yellow-headed Warbler and Red-legged Honeycreeper.

***Birds***

The following endemics, near-endemics and other specialties can be searched for: Sharp-shinned, Gundlach's and Broad-winged Hawks, Scaly-naped Pigeon, Zenaida Dove, all four species of quail-doves, Great Lizard Cuckoo, Cuban Pygmy Owl, Stygian Owl (best found in the pines around the old chalets or, even better, around the Hotel Mirador), Antillean Nighthawk (summer), Cuban Nightjar, Antillean Palm Swift, Cuban Emerald, Cuban Trogon, Cuban Tody, West Indian and Cuban Green Woodpeckers, Crescent-eyed Pewee, La Sagra's Flycatcher, Loggerhead and Giant Kingbirds (the latter rare), Cuban Vireo, Cuban Martin (summer), Cave Swallow (summer), Cuban

Solitaire, Red-legged Thrush, Olive-capped Warbler, Swainson's Warbler (winter, rare), Louisiana Waterthrush, Yellow-headed Warbler, Red-legged Honeycreeper, Western Spindalis, Cuban Grassquit, Tawny-shouldered Blackbird and Cuban Oriole.

Cuban Trogon

## Soroa

This is a tourist centre (the Villa Soroa), with a swimming pool, restaurant and bar, about 80km from Havana. At km64 of the autopista west of Havana is a gas station; turn right and drive the 7km to the hotel. In addition to the villa there are a few private houses (*casas particulares*) offering accommodation. With the exception of Blue-headed and Grey-headed Quail-Doves, and Fernandina's Flicker, nothing outstanding can be found at Soroa. The flicker can be seen on the lawn or in the woodland on the nearby hills, but is far more difficult to find here than in Zapata. A network of trails starts by the restaurant and just 100m from the grounds of the villa, along which it is possible to walk a slippery path to a well-signed viewpoint, from where you can look down on Scaly-naped Pigeons and Cuban Solitaires. Check the trails early morning and late afternoon for quail-doves; at other times of day, there may be too much disturbance. You may also find Red-legged Honeycreeper and Soroa is one of the only sites in western Cuba to find the extremely rare island race of Sharp-shinned Hawk. The grounds of the hotel harbour both Cuban Pygmy and Stygian Owls (best searched for around chalets 45–48). At the nearby stream it is possible to see Least Grebe and Louisiana Waterthrush.

***Birds***

The following endemics, near-endemics and other specialties can be found in the area: Sharp-shinned and Gundlach's Hawks, Scaly-naped Pigeon, Ruddy, Grey-headed and Blue-headed Quail-Doves, Great Lizard Cuckoo, Cuban Pygmy and Stygian Owls, Antillean Nighthawk (summer), Cuban Nightjar, Antillean Palm Swift, Cuban Emerald, Cuban Trogon, Cuban Tody, West Indian and Cuban Green Woodpeckers, Northern and Fernandina's Flickers, Crescent-eyed Pewee, La Sagra's Flycatcher, Loggerhead Kingbird, Cuban Vireo, Black-whiskered Vireo (summer), Cave Swallow (summer), Red-legged Thrush, Yellow-headed Warbler, Red-legged Honeycreeper, Western Spindalis, Cuban Bullfinch, Cuban Grassquit, Tawny-shouldered Blackbird, Eastern Meadowlark, Cuban Blackbird, Greater Antillean Grackle and Cuban Oriole.

**OTHER SITES**

## Maspoton

Close to the south coast of Pinar del Río province, Maspoton is an extensive (134km$^2$) area of lowland marshes bordered by mangrove and mudflats, which are an important stopover for migrant shorebirds. The region forms part of an Important Bird Area, which holds significant numbers of both migrant and resident waterbirds. Note that it is a very popular area with hunters and fishermen. From the city of Pinar del Río, take the autopista east towards Havana for 52km, then turn south to Los Palacios, and continue to the east end of this town, cross the railway on the right, turn immediately left and then take the first right (south) to the town of Sierra Maestra, which is reached after 11km. Then continue heading south along the main track (a heavily potholed dirt road), following signs for Club Maspoton, which offers an all-inclusive resort-style package, although some guidebooks describe it rather unfavourably (we have not stayed there). You may be better advised to day-trip the area. Almost 100 species have been recorded in the area including American White Pelican (winter), West Indian Whistling Duck, Cuban Black Hawk, Northern Caracara, Clapper Rail, 17 species of shorebirds, White-crowned Pigeon, Great Lizard Cuckoo, Antillean Nighthawk (summer), Cuban Martin (summer) and Tawny-shouldered Blackbird.

## Havana

Over 160 species have been recorded in the city limits. Die-hard birders could head for the Botanical Gardens and Havana Zoo (very close to each other); both have extensive gardens and trees that harbour both a few commoner residents and the possibility of a variety of migrants in season. There is a small charge to enter either locality, and it may be impossible to visit either early in the morning or close to sunset. A taxi from downtown Havana to either should cost US$20 each way. Near both, the Parque Lenín is north-west of the international airport and Rancho Boyeros, and represents a large green area on the edge of the city. It is very popular at weekends, and so is best visited on weekdays.

Closer to the centre of town is the Parque Almendares, in the Playa district of the city. It is easiest to get there by taxi and if staying in either the La Rampa (Vedado) or Miramar areas of the city, this should cost US$5–10 each way. Alternatively, those with their own transport should follow La Rampa away from the seafront. This avenue, number 23, eventually crosses a relatively high bridge above the Almendares River. There are parking places just off the main road on either side just beyond the bridge. Explore the parkland below, especially in the early morning for migrants. Many species of North American warblers have been recorded, including Wilson's Warbler, and other migrants have included Yellow-throated Vireo and Swainson's Thrush. Residents include the recently arrived Eurasian Collared Dove, the ubiquitous Red-legged Thrush, and Cuban Emerald, while Cuban Martin is frequent in many parts of the city from March through the summer.

At the north-east side of the city, beyond the tunnel and towards Cojímar, there is a good place for shorebirds and gulls; they gather near the shoreline in rocky, partially flooded areas. Cuba's first Franklin's Gull was found here a few years ago.

Needless to say, there are almost innumerable options for accommodation in Havana, from the relatively cheap to very expensive, with almost everything in between. Although most of those on a single-minded birding trip will probably opt to spend very little time in Havana, if you have spare days to 'kill' then there are few more special cities in the world.

## Marina Hemingway

This, the country's largest marina, lies at the western end of Havana city in the fishing settlement of Santa Fe, just beyond Jaimanitas village, with cabins, hotel and three restaurants. Marina Hemingway is reached from the capital off Avenida 5, about 5km from the tunnel on this road and is clearly signed, unlike many areas in the country. The entrance to the resort is about 1km beyond Jaimanitas, on the right. Drive to the final canal, turn left and drive on alongside the sea. Between the shore and canal there is an area with many bare rocks where, in summer, a colony of Antillean Nighthawk breeds, and a small colony of Least Terns. In winter, gulls and shorebirds can also be sought here, and Eastern Meadowlark also occurs at any season.

## San Antonio de los Baños

About 35km south-west of Havana, San Antonio de los Baños can be reached from Havana by taking the same road west out of the city, through Marianao and La Lisa, as for the autopista. The turn-off to San Antonio is well signed just before the orthopaedic hospital Frank Pais. From here it is 34km to San Antonio, where there is a low-key resort-style hotel, Las Yagrumas, just 1km north of town. The hotel is conveniently positioned within extensive riparian woods and open country along the Río Ariguanabo, with Cuban Pygmy Owl, Antillean Palm Swift, Cuban Nightjar, Cuban Green Woodpecker, Cuban Tody and Great Lizard Cuckoo. Migrants such as Yellow-throated Vireo, Blue-winged Warbler and Louisiana Waterthrush have occurred, and Common Yellowthroat is abundant.

## Laguna El Corojal

About 18km south-west of Artemisa, this lagoon is one of the best places in Cuba to find Spotted Rail, Yellow-breasted Crake and other waterfowl. Several common endemics also occur in the environs of the lake. From Havana take the autopista west for Pinar del Río and after 84km take the turn-off south to Candelaria, which is 1km from the main highway. Then, take the road through Candelaria for Mango Dulce (6km) and continue until just before the small town of Ciudad Industrial, taking the right (south) turn to El Corojal. It is about 5km from the turn to the lake. Explore the area as best you can via the roads and tracks along the west and south-west sides of the lagoon.

## Salina de Bidos

On the coast of Itabo district, about 66km east of Cárdenas, this is an excellent place for waders and waterfowl. The very rare (in Cuba) Snowy Plover occurs here, as does Cuban Black Hawk. At the opposite side of the saltpans is a small lagoon called La Lagunita, where endemics such as Oriente Warbler (at the western extremity of its range), as well as a colony of Red-shouldered Blackbird, Cuban Vireo, Cuban Green Woodpecker and Cuban Pygmy Owl also occur. To reach the area, take the road to La Teja on the coast, due north of Hoyo Colorado. The best place to stay overnight is the hotel at Baños de Elguea, north of Corralillo, or one of the hotels located on the famous Varadero beach.

# The Centre and Offshore Islands

## Isla de la Juventud

Isla de la Juventud (the Isle of Youth, formerly known as the Isle of Pines) is just 97km south of the Cuban mainland and is the second-largest island of the archipelago, covering 2,200km$^2$. It possesses both high hills and flat terrain; swamps, semi-deciduous woodland, pine forests, and savannas are all present. The southern part is mainly covered by woods on thin soils and limestone. Many of the western Cuban endemics are found here, with the exception of Cuban Parakeet, Cuban Grassquit and Cuban Blackbird.

*Location and Accommodation*

It is easiest to fly to Isla de la Juventud. The island's airport, Rafael Cabrera, is 5km from the largest town, Nueva Gerona, on the north coast, and there are either two (Monday to Thursday) or three flights per day (Friday to Sunday) from Havana. The journey takes 40 minutes each way. The other option for reaching the island is by antiquated Soviet hydrofoil, from Surgidero de Batabanó, on the south coast of the mainland. There are two sailings per day, morning and evening, and the journey takes two hours and in 2002 cost US$15. We are unaware if any birdwatchers have utilised this option and therefore whether this may be a useful means of observing seabirds, but the hydrofoil's speed is likely to be a significant negative factor in this respect. Transport around the island can be problematic, although most car drivers are willing to act as taxis, and it is possible to hire a car. There are several hotels in the north of the island, but the best resort is El Colony, about 51km south-west of Nueva Gerona, where there are also a number of *casas particulares*.

***Strategy*** At Los Indios, a sandy savanna-like area with many cabbage palms and open vegetation, there is a colony of the very rare and local Sandhill Crane. Rose-throated Parrot and woodpeckers also occur here and a breeding colony of Burrowing Owl has been found nearby. The area is a designated ecological reserve, but to date this waits approval. From El Colony, take the main road north via La Victoria to La Melvis (17km) and turn left (west) on the minor road to Mina de Oro (about 10km). Turning left (south) in Mina de Oro brings you into the open country of Los Indios. Alternatively, drive 25km south from Nueva Gerona along the island's principal highway, until you reach La Melvis.

Arguably the best birding area is the forest that covers much of the south of the island. Permission is needed to visit this area, which is a military zone. However, a permit is rather easily granted and should be sought at the tourist bureau in the Villa Gaviota hotel, on the northern outskirts of Nueva Gerona. Thereafter take the paper to the Ministerio del Interior office, on the dockside just south of Calle 16 in Nueva Gerona, to register the permit and to acquire an official stamp. You will also be obliged to take a guide (small fee), which the tourist bureau will arrange. Access is from Santa Fe (the second largest town on the island and often referred to simply as La Fe) south along the main road to Julio Antonio Mella. From there continue, following the main highway, to Cayo Piedra, 12km south of Santa Fe and passing through the military checkpoint (access will be refused without the permit), then turn towards Cuevas de Punta del Este Ecological Reserve, 23km east of Cayo Piedras. Any trail that accesses the forest off this road is likely to prove very productive. For those with more time at their disposal and a willingness to explore, the road south and then west from Cayo Piedras is also recommended. This road eventually accesses Punta Francés and the Parque Nacional Punta Francés-Punta Pedernales, about 80km from Cayo Piedras: the entire area should produce many of the more attractive endemics, such as Cuban Trogon, Cuban Tody and Cuban Green Woodpecker.

***Birds*** Over 160 species have been recorded including the following endemics and other species of interest: Least Grebe, Anhinga, Yellow-crowned Night Heron, White Ibis, Roseate Spoonbill, Wood Stork, American Flamingo, Fulvous and West Indian Whistling Ducks, Snail Kite, Northern Harrier, Cuban Black Hawk, Northern Caracara, Clapper and King Rails, Limpkin, Sandhill Crane, Northern Bobwhite, Least Tern (summer), Scaly-naped, White-crowned and Plain Pigeons, Key West, Ruddy and Blue-headed Quail-Doves, Rose-throated Parrot, Great Lizard Cuckoo, Bare-legged, Cuban Pygmy, Stygian and Burrowing Owls, Antillean Nighthawk (summer), Chuck-will's-widow (winter),

Cuban Nightjar, Antillean Palm Swift, Cuban Emerald, Bee Hummingbird, Cuban Trogon, Cuban Tody, West Indian and Cuban Green Woodpeckers, Crescent-eyed Pewee, La Sagra's Flycatcher, Grey, Loggerhead and Giant Kingbirds, Cuban Vireo, Black-whiskered Vireo (summer), Cuban Crow, Cuban Martin (summer), Cave Swallow (summer), Red-legged Thrush, Bananaquit (vagrant), Yellow-headed Warbler, Western Spindalis, Cuban Bullfinch, Red-shouldered Blackbird, Eastern Meadowlark, Greater Antillean Grackle and Cuban Oriole.

## Topes de Collantes

About 12km north of the city of Trinidad (one of the five oldest cities in Cuba and a World Heritage Site), and 750m above sea level, the Topes de Collantes lie within the 110km$^2$ of the Sierra de Escambray and in a Natural Park (small entry fee). The main habitats of this Important Bird Area are submontane seasonal rainforest, tropical karstic forest, pine forest and gallery forest.

***Location*** From Trinidad drive west along the main highway towards Cienfuegos for 4km and take the well-signed right (north) turn to the Topes. The hotels are 14km along this road, and it is possible to explore some of the highland forests by road, either by taking that to Jibacoa (3km beyond the hotels) or to La Sierrita (take the left fork where the road to Jibacoa lies to the right).

***Accommodation*** There are three hotels high up at Topes de Collantes, from where a network of trails accesses the surrounding forest. It is also possible to arrange day trips to the Topes with one of the usual tour agencies, Cubatur, Cubancán or Rumbos, in Trinidad, but there is no public transport to the area.

***Strategy*** Among the most interesting species present in the Topes are Gundlach's and Sharp-shinned Hawks, Scaly-naped Pigeon, Cuban Parakeet, Rose-throated Parrot, Cuban Pygmy, Bare-legged and Stygian Owls, White-collared and Black Swifts (this is one of the best areas in the western half of the island for both swifts), Cuban Trogon, Cuban Tody, most of Cuba's woodpecker species, Cuban Vireo, Western Spindalis and Cuban Bullfinch.

By continuing west along the main road to Cienfuegos, you can access some reasonably interesting areas of scrub and partially wooded habitats where it is sometimes possible to find the globally threatened Giant Kingbird and increasingly rare Cuban Grassquit. In addition, only 7km south of Trinidad is Casilda, a fishing village, near which several waterbirds occur. Baird's Sandpiper, a species not yet certainly recorded in Cuba, was claimed here recently. On the small peninsula to the south there is a nice beach, the Playa de Ancón, and several reasonable hotels, close to which it should be possible to locate Cuban Gnatcatcher in the surrounding scrub.

***Birds*** Over 100 species have been recorded in this area including the following species of particular interest to the visiting birder: Least Grebe, Yellow-crowned Night Heron, Gundlach's, Sharp-shinned and Red-tailed Hawks, Northern Caracara, Northern Bobwhite, King Rail, Limpkin, Northern Jacana, Scaly-naped and White-crowned Pigeons, Zenaida Dove, Key West, Ruddy and Blue-headed Quail-Doves, Rose-throated Parrot, Great Lizard Cuckoo, Bare-legged, Cuban Pygmy and Stygian Owls, Antillean Nighthawk (summer), Cuban Nightjar, Black, White-collared and Antillean Palm Swifts, Cuban Emerald, Cuban Trogon, Cuban Tody, West Indian and Cuban Green Woodpeckers, Northern Flicker, Crescent-eyed Pewee, Grey, Loggerhead and Giant Kingbirds, Cuban Vireo, Cuban Crow, Cuban Martin (summer), Cave Swallow (summer), Cuban Gnatcatcher, Red-legged Thrush, Red-legged Honeycreeper, Western Spindalis, Cuban Bullfinch, Cuban Grassquit (rare), Tawny-shouldered Blackbird, Eastern Meadowlark, Cuban Blackbird, Greater Antillean Grackle and Cuban Oriole.

## Cayo Coco

This, the second-largest key in Cuba, is connected to the mainland by a rock-fill road and is about 70km north-west of Morón. The island is mostly covered by woodland, but there are also mangroves, coastal shrubs, grassy areas and lagoons. Over 200 species have been reported, including many Cuban rarities and several new birds to the entire West

***Map of Cayo Coco***

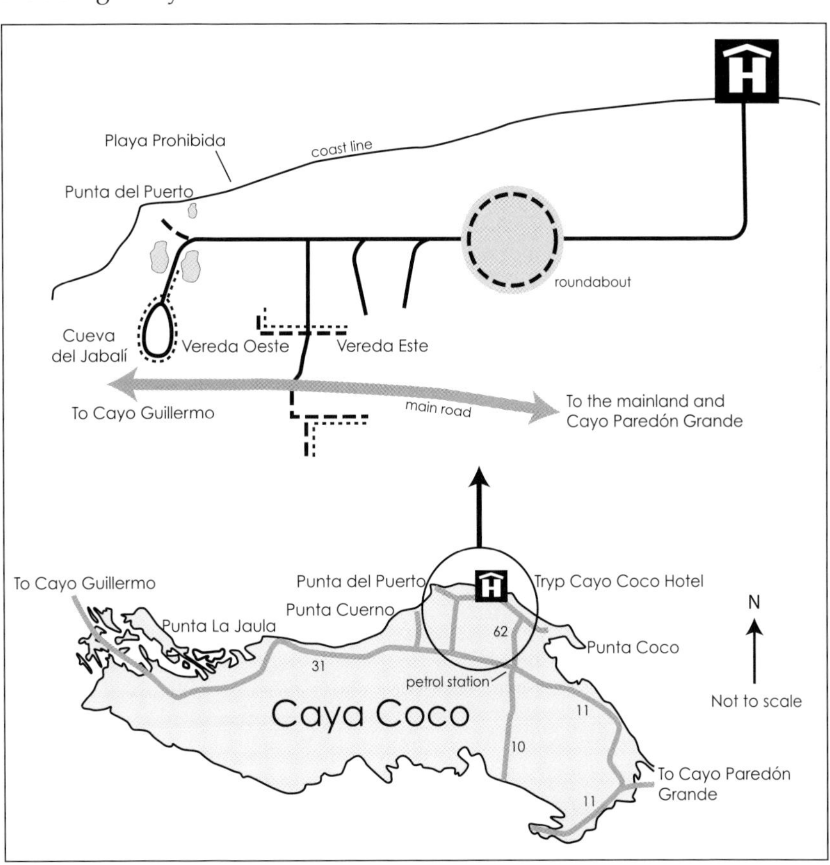

Indies (for instance Black-throated Grey Warbler), although the most important and interesting species to visiting birders are Ruddy and Key West Quail-Doves, a local (greyer) race of Great Lizard Cuckoo (*santamariae*), Cuban Tody, West Indian and Cuban Green Woodpeckers, Northern Flicker, Cuban Gnatcatcher, the *varonai* race of Zapata Sparrow, Western Spindalis and Cuban Bullfinch. There are many waders and one of the largest groups of American Flamingo known in the country. In autumn, Merlin and Peregrine Falcon are not uncommon, and this is also the best period for passerine migrants and vagrants.

*Accommodation*

There are several large hotel complexes of the all-inclusive type on the north side of the cay. Many birders may regard these with something akin to horror and too expensive for their taste. Truth is, however, that they are comparatively inexpensive for what they offer and they are certainly much more convenient than staying on the mainland, in Morón, or using one of the cheap bungalows at Flamingo Beach or Sitio La Güira, neither of which we have personal experience, but are apparently not recommended. Note that it is approximately 45 minutes each way between Morón and Cayo Coco and there is a $3 toll per car/per day (at which passports are also checked) if you drive out from the mainland each day.

*Strategy*

From the Club Tryp Hotel, or any of the other large hotels on the northern shore of Cayo Coco, turn right and drive for about 5km passing through a roundabout. At a bend in the road (2km from the hotel) there is dirt road to the right (below a sign 'Cueva el Jabalí'). Continue along this road for 3km to the intersection with two dirt roads, both of which are good for the Zapata Sparrow and Oriente Warbler. Another interesting location is reached by driving straight at the sign for the Jabalí to the end; there is a parking lot after about 3km where the cave is located. Surrounding the area is a trail that enters the forest, where Key West Quail-Dove is often not hard to see, especially around the drinking pool at the end, in early morning or late afternoon (during the dry season). This area is also good for Oriente Warbler, Cuban Green Woodpecker and several migrants, occasionally including rarities such as Black-billed Cuckoo. Another area worth checking is a series of ponds on the left (800m before the cave) just 100m before a right-hand bend in the road with the ocean in front of you. Just left of the bend there is a sandy track, when you reach the end (on the beach) turn left and walk along the shore for 40m, and then turn left onto a narrow sandy track through the mangrove which will soon bring you to a pond.

Another productive pond can be reached by taking the main road to Cayo Guillermo, and turning left before you reach the sign for Flamingo Beach; this track accesses a wetland with many ducks in season, including West Indian Whistling Duck (which can also be found in the mangrove-lined lagoons around some of the hotels). Any trail that takes you to the forest or into mangrove is good for birding. One, which we have not tried but has been recommended, is located in undisturbed forest, about 1km south of the petrol station at the first roundabout on the southern side of the cay.

North of Cayo Coco there are three small rocky keys, called Los Felipes, where several species of terns breed in the summer. Anyone birding Cayo Coco should visit the two adjacent cays, Cayo Guillermo to the west and Cayo Paredón Grande to the east, both of which harbour some additional species that are more easily found there (see Other sites below).

***Birds***

In addition to the migrants and vagrants, the following regularly occurring species are of particular interest: many herons and egrets, Roseate Spoonbill, Wood Stork, American Flamingo (best looked for from the causeway, although views are typically rather distant), West Indian Whistling Duck, many ducks including large numbers of Red-breasted Merganser (for many years this was the only regular wintering locality in the Caribbean), Gundlach's Hawk (rare), Cuban Black Hawk (one of the best localities in Cuba), Northern Caracara, Clapper Rail, many waders including the globally Near Threatened Piping Plover (best looked for at Flamingo Beach), Least Tern (summer), White-crowned Pigeon, Zenaida Dove, Key West and Ruddy Quail-Doves, Mangrove (uncommon) and Great Lizard Cuckoos, Bare-legged Owl, Chuck-will's-widow (winter), Cuban Nightjar, Cuban Emerald, Cuban Tody, West Indian and Cuban Green Woodpeckers, Northern Flicker, Crescent-eyed Pewee, La Sagra's Flycatcher, Loggerhead Kingbird, Thick-billed Vireo (probably only a vagrant), Cuban Vireo, Black-whiskered Vireo (summer), Cuban Martin (summer), Cave Swallow (summer), Cuban Gnatcatcher, Red-legged Thrush, 32 species of wood warblers, including Oriente Warbler, Western Spindalis, Cuban Bullfinch, Zapata Sparrow, Tawny-shouldered Blackbird, Greater Antillean Grackle and Cuban Oriole.

## Sierra de Najasa

Located about 70km south-east of Camagüey, this is a protected area of open country with many palm groves, semi-deciduous woods, and tropical karstic forest in the hills. About 120 species of birds have been reported, among them Plain Pigeon, Cuban Parakeet, Rose-throated Parrot, Fernandina's Flicker and Giant Kingbird. In two or three areas north, south and east of Camagüey (Miguel, El Jardín, Tayabito) the very rare and local Cuban Palm Crow can be found.

***Location and Strategy***

To reach Najasa, drive south from Camagüey city, taking a road known as Carretera de Santa Cruz, which can be accessed by turning left from the Hotel Camagüey and left again near a gas station (the latter is signed). Drive south for approximately 43km. After 30km the road conditions deteriorate significantly and it will be necessary to reduce your speed. Turn right at km43, by a bus stop, following a sign to Vertientes. Drive a further 22km to the village of Cuatro Caminos (Najasa on some maps), stopping just 1km beyond the Carretera de Santa Cruz at a large area of palms on both sides of the road, where Plain Pigeon, Cuban Parakeet, Cuban Palm Crow and Cuban Crow are all possible. Upon arriving at Cuatro Caminos, turn right at the first

*Map of Sierra de Najasa*

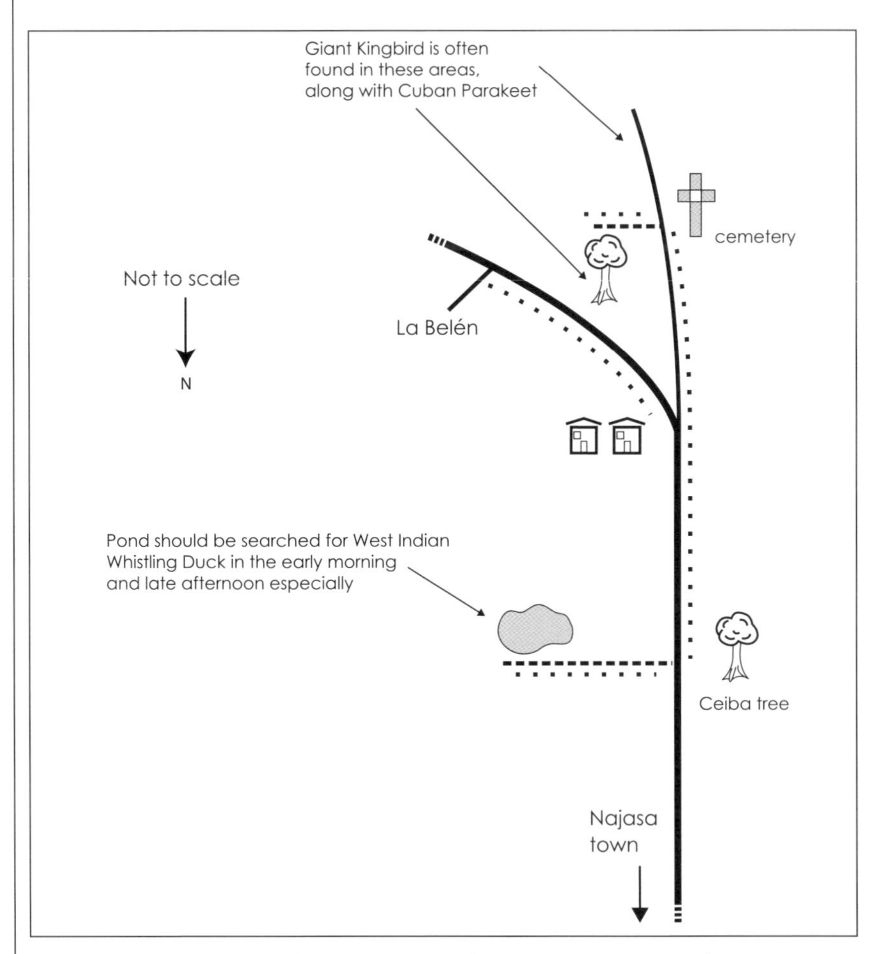


intersection (by a cup-shaped cafeteria) and drive for a further 8km until you reach the village of Arroyo Hondo (also known as Reparto Manolin). Shortly after reaching the village there is a junction. Good birding can be had in this area. The Najasa reserve area (La Belén) can be reached by turning left and continuing for 1km; the entrance is immediately beyond the row of houses on the left-hand side of the road. Further along the road beyond the reserve entrance there is a pond where large numbers of West Indian Whistling Ducks sometimes come into roost early in the morning (like other whistling ducks the species is mainly active nocturnally). By following the road to the right, you reach a small cemetery on the right-hand side. This area can be very productive for Cuban Parakeet and Giant Kingbird.

When returning to Cuatro Caminos, turn left towards Carretera de Santa Cruz. Approximately 5km before you reach the latter village there is a small group of houses; on the right is a ruin with a star monument. Take the road to the right of this, which accesses a dam, where it is possible to observe a variety of waterbirds. Also, 3km south of Cuatro Caminos village, on the right-hand (east) side of the road there is a large *Ceiba* tree with an open yellow triangle painted on its base and a dirt track on the same side of the road immediately before it; 10m beyond this, on the left-hand side of the road, there is a trail with a fence across its entrance. Walk about 600m through quite dense scrub to a pond on the right-hand side where West Indian Whistling Duck can be found roosting late in the afternoon or early in the morning on the first large rotten tree or elsewhere on the lake. Masked Duck has also been recorded here.

Another way (perhaps more straightforward) to reach Najasa is to drive east on the Carretera Central towards Las Tunas province. From the Hotel Camagüey turn right onto the main road and continue 13km beyond the sign marking the Camagüey city limits and, immediately beyond a Cupet gas station on a small hill, turn right to Cuatro Caminos (27km), and continue as described above to reach Arroyo Hondo. Barn Owl can sometimes be found around the gas station at night, as well as in the La Belén reserve. Gundlach's Hawk, Plain Pigeon, Cuban Palm and Cuban Crows, Giant Kingbird, Tawny-shouldered Blackbird and Eastern Meadowlark can all be found along the road between the Carretera Central and Cuatro Caminos. There are also a few small wetland areas along this road, which can be worth checking for those with more time.

***Accommodation***

There are several regular hotel options in the city of Camagüey, of which the Gran is perhaps the best, and the Hotel Camagüey, which has had a facelift in recent years, has the advantage, for anyone self-driving, of being on the edge of town and well sited for the birding, as well as being much less difficult to find. The centre of the city is a near-maze to the uninitiated, and not easy to find your way around, even with experience. There are also plenty of *casa particulares* to choose from; as always, Cubans will approach you offering such possibilities, although many of the better private houses are now listed in the usual travel guides.

***Birds***

The following species are perhaps of greatest interest in this area: Least Grebe, Anhinga, West Indian Whistling, Wood and Masked Ducks, Snail Kite, Sharp-shinned and Gundlach's Hawks (neither is regularly seen), Northern Caracara, Limpkin (surprisingly common in this generally dry country), Northern Bobwhite, Northern Jacana, Scaly-naped, White-crowned and Plain Pigeons (the latter in one of its Cuban strongholds), Ruddy and Blue-headed Quail-Doves (both very hard to see in this region), Cuban Parakeet, Rose-throated Parrot, Great Lizard Cuckoo, Barn Owl, Cuban Pygmy and Bare-legged Owls, Cuban Nightjar, Antillean Palm Swift, Cuban Emerald, Cuban Trogon, Cuban Tody, West Indian and Cuban Green Woodpeckers, Northern and Fernandina's Flickers, Crescent-eyed Pewee, La Sagra's Flycatcher, Grey, Loggerhead and Giant Kingbirds (the area is one of the island's major strongholds for the latter species), Cuban Vireo, Black-whiskered Vireo (summer), Cuban Palm Crow (the best site for the species), Cuban Crow, Cuban Martin (summer; common in the city), Cave Swallow (summer), Red-legged Thrush, Western Spindalis, Cuban Bullfinch, Cuban Grassquit (uncommon), Tawny-shouldered Blackbird, Eastern Meadowlark, Cuban Blackbird, Greater Antillean Grackle and Cuban Oriole.

**OTHER SITES**

## Sierra de Cubitas

The sierra is about 30km north-east of the city of Camagüey. From the centre of the latter take the road to the city's international airport. Directly in front of the main terminal building turn left by a bus stop, cross the railway line, and follow the road north for 24km. Just beyond a large sign, announcing the Sierra de Cubitas, the road forks. Go right and continue along a dirt road for about 6km beyond the end of the paved section. Just beyond a house with a windmill you should turn left and drive 500m. Walk the area between the two cliffs within a low-elevation area covered by deciduous woodland. Many birds can be found here: Cuban Pygmy Owl, Cuban Trogon, Cuban Tody, West Indian and Cuban Green Woodpeckers, Oriente Warbler and a variety of migrants. Before reaching this area, you will pass by Paso de Lesca, an open shrubby savanna-like area with many cabbage palms and xerophytic vegetation, which is a good place to find Cuban Vireo and Oriente Warbler.

To reach Cayo Coco from the Sierra de Cubitas (the area makes a good stop-off en route between Camagüey and the northern cays), return to the junction just beyond the sign and take the left-hand fork. Follow this road to the next intersection, turn left and continue through the town of Esmeralda, which is several kilometres beyond the junction. Continue through the town of Morón, forking right at a large junction with a sign to Cayo Coco and an Oro Negro petrol station on one corner. The drive from Camagüey to the cays takes about 3 hours via the sierra.

Before reaching the junction with the Oro Negro station it may be worth checking the grazing marshes east of the small village of Bolivia. To reach the best and most reliable area in this part of Cuba for the endemic race (*nesiotes*) of Sandhill Crane take the track south from the main road 24.5km east of the intersection and drive for 5.4km before taking a rough track west (right) into an extensive marsh and grazing area, where Sandhill Crane may be regular, although as usual early mornings are best and visits later in the day are often unsuccessful. Gundlach's Hawk has also been seen here recently, and the marshes and lagoons further west, closer to Morón, may be even more productive for a range of wetland species and waterbirds, especially in winter and during passage periods.

## Cayo Cantiles

The second-largest cay off the southern coast, 48km west of Cayo Largo del Sur, it has semi-deciduous woodland, second growth, evergreen forest (Key West Quail-Dove), mangrove and lagoons (West Indian Whistling Duck). Almost 60 species have been reported, including, Cuban Green Woodpecker (the population here was formerly accorded subspecific status, '*gloriae*'), a race of Cuban Vireo (*magnus*), Yellow-headed Warbler, and a race of Tawny-shouldered Blackbird (*scopulus*). Mangrove Cuckoo is not uncommon here. During spring migration, several warbler species can be expected. The cay can be reached only by boat from Cayo Largo del Sur, where there are, at present, at least seven hotels.

## Cayo Largo del Sur

Larger than Cantiles, this largely coral island possesses much less varied vegetation than the latter, and therefore a slightly lesser diversity of birds. It is about 26km long but less than 2km wide. Common birds include Antillean Nighthawk (summer), Cuban Black Hawk and Osprey. There is a muddy and sandy area in the west of the cay, where during autumn migration several species of shorebirds gather. Just to the south-west is Cayo Los Ballenatos, a pair of rocky cays where several species of terns (including Bridled, Sooty and Least), and Laughing Gull, breed. There are at least seven all-inclusive (and therefore not cheap) hotels on the cay, which is accessible by both charter and regular flights from Havana (it is even possible to day-trip the island from the capital), with flights also available from Varadero and Isla de la Juventud. There is also a boat service from the latter. It is possible to hire a car to explore the cay, as two agencies have outlets at the Hotel Pelicano, and motorcycles and bicycles are also available to rent.

## Cayo Tío Pepe

Located north-east of Isabela la Sagua, in Villa Clara province, this island represents the only known Cuban breeding grounds of Black-faced Grassquit. Bahama Mockingbird is also present, as is Antillean Nighthawk (summer), Crescent-eyed Pewee and Cuban Green Woodpecker, and a specimen of Bananaquit has been collected on the cay. However, there is no accommodation on the island, and the nearest hotel is probably in Sagua la Grande, about 20km inland, unless you can find a *casa particular* in Isabela la Sagua.

## Cayo Paredón Grande

From Cayo Coco, you can reach Cayo Paredón Grande by turning left at the entrance to the Club Tryp Hotel and driving along the main paved road. Go straight ahead at the first roundabout and then turn left at the second roundabout, and pass through a built-up area. Turn left onto the wider road and keep straight ahead to a junction, where you turn left and continue straight ahead again to the cay. Your destination will soon become visible in the distance due to its black-and-yellow painted lighthouse. At the point where you cross a rather narrow bridge between Cayo Coco and Cayo Romano, there may be barrier across the road. Ask the workers in the building on the left to open it for you. They might ask what you are doing. Park by the lighthouse (37km from the hotels on Cayo Coco) and walk along the trail to the right as far as the beach (1km), where the mangroves just inland harbour Clapper Rail. Almost 125 species have been reported on this small island including

Piping Plover (check the beach) and Thick-billed Vireo (of the recently described endemic subspecies *cubensis*). The lighthouse may attract large numbers of North American warblers, especially in autumn, and the trail can also prove productive for migrants. The potential for vagrants is high on this isolated cay; several scarce migrants have been reported, and, in April 2001, the first Bahama Woodstar on Cuba was claimed here. It is also an excellent area for Mangrove Cuckoo, Cuban Green Woodpecker, Bahama Mockingbird, Cuban Gnatcatcher, Oriente Warbler (the two latter common and easily located, especially with a tape) and, among seabirds, Brown Booby, with various terns possible in summer.

## Cayo Guillermo

Situated 34km north-west of the hotels on Cayo Coco is Cayo Guillermo, which is reached by turning left at the first large roundabout on Cayo Coco (with the petrol station) and driving straight until you cross a relatively long bridge with a series of hotels on the beach to the right; this is Cayo Guillermo. The area around the bridge is a good for gulls and terns, and, at low tide, shorebirds, which may include some of the scarcer species. The pools lining the road in front of the hotels are also very productive at times for shorebirds (including many Stilt Sandpipers) and other large wading birds, as well as Clapper Rail. It is a sandy cay with xerophytic vegetation and many palms *Coccothrinax* sp., haunt of Bahama Mockingbird, which is the key target for visiting birders. It is necessary to drive along several kilometres of the main gravel road beyond the hotels to reach the best area for the mockingbird, which is being out-competed by the recently arrived Northern Mockingbird, and has also declined due to the advent of tourism on the cay. Bahama Mockingbirds are almost invariably found in pairs, the two birds often in very close proximity. The area around a dirt track off to the right, about 4km beyond the hotels, usually holds several pairs, and Thick-billed Vireo, of the endemic Cuban subspecies, has also been found here on a few occasions in recent years. While most birders choose to stay on Cayo Coco, or even in Morón, there are several hotels on the cay itself, all offering similar-standard (and priced) accommodation to that available on Cayo Coco.

## Cayo Romano

The largest of the Cuban cays and, like Cayo Coco, connected to the mainland by a 20km rock-fill causeway. The vegetation is also similar to that on Cayo Coco, and there is a road from the centre of the island to the extreme eastern end, Punta El Inglés, and another leading north that eventually proceeds via another causeway to Cayo Cruz. There are no hotels on the cay, and the nearest regular accommodation may be as far away as Nuevitas or even Camagüey, unless one can find a *casa particular* in Esmeralda or nearer. Some 130 species have been reported including Zapata Sparrow and most of the eastern endemics, as well as many migrants. In Cuba, Northern Caracara is generally rather rare, but on this cay it is, by contrast, quite common.

## Cayo Cruz

This is a long narrow cay, about 10km north of Cayo Romano to which it is linked by a rock-fill causeway, with habitat quite similar to Cayos Paredón Grande and Guillermo, including many palms, *Coccothrinax*. Cayo Cruz harbours the largest Cuban population of Bahama Mockingbird, but like Cayo Romano is not served by any hotels. Just over 50 species have been recorded to date including a few endemics, e.g. Cuban Green Woodpecker and Oriente Warbler. Accommodation options are described under the previous site.

# The Oriente

## Sierra de Nipe-Sagua-Baracoa, Mayarí

This mountain range in north-east Cuba borders the Atlantic on its eastern and northern peripheries, the Central and Guantánamo Valleys to the south and, in the west, the plains of Nipe and Cauto. The highest peak, at 1,231m, is Pico Cristal and the range stretches almost 190km west to east. Regionally, the area is a notable watershed, the following rivers having their sources within the sierra: Toa, Yumurí, Duaba, Moa, Jobo and Imias, among others. Habitats include large areas of native pine forest, as well as a notable diversity of other forest types, namely second-growth forest, evergreen forest, gallery forest, montane tropical rainforest, semi-deciduous forest and submontane seasonal rainforest, tropical karstic forest, montane serpentine shrubwoods, and pine forest.

***Location***

The via Mulata (a road) runs north-east from Guantánamo to Baracoa, thus traversing and permitting access to much of this range, via Palenque, Bernardo, Las Calabazas and Quibiján. Possibilities along this road include Gundlach's Hawk, Rose-throated Parrot, Cuban Parakeet, White-collared Swift, Stygian and Bare-legged Owls, Cuban Trogon, Cuban Solitaire and Oriente Warbler. The area also appears suitable for the extremely poorly known Cuban Kite, which has been reported on several occasions from this area by local farmers and its favoured land snail prey has been discovered to be common in parts of the region. It may also be possible to access this mountain range by driving north from San Antonio del Sur to Puriales de Caujerí and thence, via a dirt road unmarked on most maps, to the via Mulata between Bernardo and Las Calabazas.

***Accommodation***

The nearest towns are Baracoa, Mayarí, Sagua de Tánamo, Imias, San Antonio del Sur and Cajobabo, of which we recommend the first as a base for exploring this range, although Moa is perhaps closer but less well served with tourist facilities.

***Strategy***

For those with limited time at their disposal, we recommend visiting La Melba, about 44km by road south-southeast of Moa. From the latter proceed south-east along the coast towards Baracoa and turn inland (right) on a dirt road just after crossing the Río Guam, about 4km beyond Punta Gorda and almost 13km from Moa, and just in front of an old mine. After 1km, take the left turn and continue on the dirt road for about 30km to the small village of La Melba. Just south-east of the settlement there is a small stream (Arroyo Sucio) with gallery forest around which Cuban Kite has been reported by local people on several occasions. Other interesting species to be found here include Sharp-shinned and Gundlach's Hawks, Rose-throated Parrot, Cuban Parakeet, three owls, Cuban Trogon, Giant Kingbird, Cuban Solitaire, Cuban Grassquit and many others.

***Birds***

More than 100 species have been recorded in this area, despite a relative lack of birdwatching activity to date. These include: Least Grebe, Snail Kite, Sharp-shinned and Gundlach's Hawks, Northern Caracara, Northern Bobwhite, Limpkin, Scaly-naped Pigeon, White-winged and Zenaida Doves, Ruddy and Grey-headed Quail-Doves, Cuban Parakeet, Rose-throated Parrot, Great Lizard Cuckoo, Cuban Pygmy, Bare-legged and Stygian Owls, Antillean Nighthawk (summer), Chuck-will's-widow (winter), Cuban Nightjar, White-collared and Antillean Palm Swifts, Cuban Emerald, Bee Hummingbird, Cuban Trogon, Cuban Tody, West Indian and Cuban Green Woodpeckers, Northern Flicker, Crescent-eyed Pewee, La Sagra's Flycatcher, Loggerhead and Giant Kingbirds, Cuban Vireo, Black-whiskered Vireo (summer), Cuban Crow, Cave Swallow (summer), Cuban Solitaire, Red-legged Thrush, Olive-capped and Oriente Warblers, Red-legged Honeycreeper, Western Spindalis, Cuban Bullfinch, Cuban Grassquit, Tawny-shouldered Blackbird, Eastern Meadowlark, Cuban Blackbird, Greater Antillean Grackle, Cuban Oriole and Nutmeg Mannikin.

## Birama Swamp

Chiefly interesting for the large number of aquatic birds, these include a breeding colony of up to 30,000 American Flamingos and mixed colonies of ibis and herons totalling 15,000 pairs, as well as the globally threatened West Indian Whistling Duck. The Cauto Delta, as a whole, hosts up to 100,000 waterbirds in winter, which has led to it being designated an Important Bird Area. In addition, a number of woodland endemics are also known from the area, including the globally threatened Gundlach's Hawk and Cuban Parakeet, and the region is considered a stronghold of the equally rare Fernandina's Flicker. Typical habitats of this Ramsar site are swampy forest, mangrove, semi-deciduous woodland and savanna. Two wildlife refuges have been established, but illegal hunting and fishing, fires and timber extraction remain ongoing threats.

***Location***

This area, in Granma province, is one of the most important wetlands in Cuba and the entire Caribbean, sited in the Cauto plains, adjacent to the Gulf of Guacanayabo, and 90km north-west of Manzanillo and south of Jobabo.

***Accommodation***

There is a small hotel at Biramas (reached by turning left off the road about 10km north of Guamo Embarcadero or via a more direct dirt road from Guamo Embarcadero itself) on the north-east side of the Embalse Leonero which, along with the Laguna Birama (inaccessible by road) in the west of the area, is the best area for wildfowl.

***Strategy***

Much of the area is largely inaccessible without a boat, although these may be hired, along with a guide, from near the large bridge in Guamo Embarcadero on the Río Cauto (which at 370km is Cuba's longest river). It is also possible to hire a boat and guide from Cauto Embarcadero, just off the main road, 52km south of Las Tunas, but this will necessitate a longer journey before reaching the most interesting birding areas. Although difficult to access, the area is interesting year-round, with visits during passage periods and winter having revealed spectacular concentrations of birds, sometimes of species considered rare in Cuba, while summer is dominated by the large breeding colonies.

West Indian Whistling Duck

***Birds***

Over 100 species have been recorded during recent surveys; these include: Least and Pied-billed Grebes, Anhinga, many herons and ibis, American Flamingo, West Indian and Fulvous Whistling Ducks, Wood and Masked Ducks, Osprey, Snail Kite, Northern Harrier, Gundlach's and Cuban Black Hawks, King, Clapper and Spotted Rails, Purple Gallinule, Limpkin, American Avocet (winter), Wilson's Plover, Gull-billed Tern (sometimes large numbers in winter), White-crowned Pigeon, Cuban Parakeet, Mangrove and Great Lizard Cuckoos, Cuban Pygmy Owl, Antillean Nighthawk (summer), Cuban Emerald, Cuban Tody, West Indian and Cuban Green Woodpeckers, Fernandina's Flicker, Crescent-eyed Pewee, La Sagra's Flycatcher, Cuban Vireo, Black-whiskered Vireo (summer), Cuban Martin (summer), Red-legged Thrush, Western Spindalis, Cuban Bullfinch, Cuban and Tawny-shouldered Blackbirds.

## Baitiquirí

This xerophytic area on the south coast, which is the driest in all of Cuba, is particularly notable for the local race of Zapata Sparrow (*sigmani*) which is apparently restricted to a tiny area of littoral between here and the village of Imías, 25km further east. The population has recently been estimated at just 600–700 birds. In contrast to the other two forms, which are relatively similar in morphology, *sigmani* is considerably duller, the crown being almost grey. Thus far, no DNA studies have been conducted of the three races, although comparison of the vocalisations of *sigmani* and nominate *inexpectata* has revealed few differences. What is clear, however, is that *sigmani* is rather uncommon and finding this interesting form will require a certain amount of dedication, although recent surveys have revealed a larger and more widespread population than previously thought.

***Location and Accommodation***

The main birding area is about 50km east of the city of Guantánamo, in which you will find the nearest hotel for exploring the region.

***Strategy***

The vegetation in this area, a recently designated Important Bird Area, is characterised by rather dense, coastal thickets, cactus scrub with tall, columnar cacti and some scattered trees, and dry evergreen forest. Common in such areas are Cuban Vireo and Oriente Warbler, while Cuban Grassquit can be positively abundant. Much less common and only present in winter, when the species descends from the mountains, your chances of finding a Bee Hummingbird in the Baitiquirí region will require some persistence and good fortune. Northern Potoo was recently discovered here, which is reminiscent of those areas in which it is most common and easily found in Hispaniola, while Cuban Nightjar is rather common along the road between here and Tortuguilla after nightfall. There is a small colony of White-tailed Tropicbird in the region and the birds may be seen offshore during the breeding season (from March until early summer).

One of the best areas of dry scrub is that about 2.5km west of the village, at the point where the main coastal road cuts inland through a small valley dissecting the low ridge that runs parallel to the coast here. Immediately beyond the point where the road curves to the left, take a narrow dirt track off to the right. Park after a short distance (out of sight of the road) and explore the tall scrub anywhere in the area, which is perhaps the best for the Zapata Sparrow. It is possible to walk a couple of kilometres along the coast here, with Cuban Grassquit and Cuban Gnatcatcher among the commonest birds. However, for those with time at their disposal there are many other areas of identical and equally suitable habitat in the general area. The village overlooks a natural harbour, where Brown Pelican and other coastal species are often present, backed by a small area of saltpans, which lie immediately adjacent to the main road at the east end of Baitiquirí. Given favourable water levels these can be productive for waders and herons, and will be particularly worthy of scrutiny in passage periods.

Continuing about 10km further east along the main road there is reasonable-sized, shallow wetland, again on the right-hand side of the road, as you draw near the town of Macombo. This area is also very much worth a look, with the potential for interesting shorebirds and waterfowl, especially during passage seasons, and plenty of passerines in the surrounding scrub, including both Cuban Gnatcatcher and Oriente Warbler, and a reasonable array of overwintering Nearctic warblers in season. Chances for an interesting migrant or two appear very favourable. Note that wellingtons or other strong footwear may be advisable at this locality, and that it is necessary to traverse a barbed-wire fence to get close views of the pond. West Indian Whistling Duck has recently been recorded in mangroves near the town of San Antonio del Sur, between Baitiquirí and Macambo.

***Birds*** Just under 100 species have been recorded to date, but this list is still based on rather few visits, and we consider the area to be one of the most fascinating and potentially most rewarding birding destinations in eastern Cuba, especially given its relative accessibility. Species of particular interest include: White-tailed Tropicbird, West Indian Whistling Duck, Wilson's Plover, Northern Jacana, Scaly-naped and White-crowned Pigeons, White-winged Dove, Great Lizard Cuckoo, Cuban Pygmy Owl, Northern Potoo (one of the few sites in Cuba), Antillean Nighthawk (summer), Chuck-will's-widow (winter), Cuban Nightjar (common along roads), Black and White-collared Swifts, Cuban Emerald, Bee Hummingbird, Cuban Tody, Crescent-eyed Pewee, La Sagra's Flycatcher, Loggerhead Kingbird, Cuban Vireo, Black-whiskered Vireo (summer), Cuban Martin (summer), Cave Swallow (summer), Cuban Gnatcatcher, Oriente Warbler, Cuban Bullfinch, Cuban Grassquit (common), Zapata Sparrow, Eastern Meadowlark and Tawny-shouldered Blackbird.

## Embalse La Yaya

We have only visited this area once, in January 2001 (when water levels appeared to be rather low), but have included it here on the basis of its potential and accessibility. Especially in winter and during passage periods, it is clearly worth checking for waterfowl and shorebirds.

***Location*** This quite large, shallow reservoir is about 15km from the town of Guantánamo, on the south side of the main road to Santiago de Cuba. If approaching from the latter, it is on the right-hand side of the highway.

***Accommodation*** There is a reasonable variety of accommodations in Santiago de Cuba, but just one hotel in Guantánamo.

***Strategy*** Trees and scrub screen much of the western end of the reservoir, but approximately halfway along its length a short track cuts downslope through an area of woodland. Park by a crossroads in the track; the reservoir can be reached by walking either straight ahead and then through the tall grass on the north shore, or by walking left through the woodland and across open ground to the north-east arm of the dam. From here, by walking right, along the shore of the reservoir, it would be possible to return to your vehicle via a roughly circular route. It may also be possible to view the extreme eastern end of the reservoir by continuing along the main highway to the first paved turning to the right, and then shortly afterwards right again on the dirt road to Vilorio. In addition, do not ignore the surrounding woodland, which can be productive for North American migrants. The reservoir makes a convenient stop for an hour or so if travelling between Santiago and Guantánamo.

***Birds***

Some of the more interesting species recorded during our only visit included: Anhinga, Glossy Ibis, White-cheeked Pintail, Limpkin, Caspian Tern, White-winged Dove, Great Lizard Cuckoo, White-collared Swift, Cuban Emerald, Cuban Trogon, Cuban Tody, Belted Kingfisher, Crescent-eyed Pewee, Cuban Vireo, Cuban Gnatcatcher, Red-legged Thrush, Cuban Grassquit, Eastern Meadowlark and Cuban Blackbird.

**OTHER SITES**

## La Silla de Gibara

Situated in the highlands of Maniabón, 7km south of the town of Potrerillo and 34km north-east of Holguín, this area of karstic forest and dry evergreen forest reaches a maximum altitude of just over 300m. From Holguín, take the road to Gibara and after 16km, just before the small town of Aura, turn right and drive another 13km along a dirt road to the Silla. There is a resort with chalets and it is also possible to stay in nearby Gibara (several *casas particulares*), from where it is even possible to access the Silla via a cable-car network, or Holguín. The area is also within easy striking distance of the coastal resort of Guardalavaca, which lies slightly further east. From the parking area by the office at the resort, follow a 2-km trail to the left towards the Silla (and an overlook) through forest where 12 endemics, including Gundlach's Hawk, Grey-headed Quail-Dove, Cuban Pygmy Owl, Cuban Nightjar, Cuban Trogon, Cuban Green Woodpecker, Oriente Warbler and Cuban Grassquit, can be found. There is also a side trail through a cavern system. The nearby lowlands support species such as Giant Kingbird, while the coastal mangroves and lagoons north of Holguín and west of Gibara host West Indian Whistling Duck, Piping Plover and White-crowned Pigeon. The whole area can be productive for migrants, including a good smattering of scarcities.

## Pinares de Mayarí

Situated 25km south of the town of Mayarí (several hotels) on the north-east coast, the resort of Pinares de Mayarí lies within the Parque Nacional La Mensura and is easily accessed by road. There is a research station operated by the Cuban Academy of Sciences in the park. Two major rivers flow through the area, the Nipe and Mayarí. Fifty-four per cent of the park is still forested, with semi-deciduous woodland, montane serpentine shrub woodland, tropical karsitic forest and much pine forest. Good birding is possible in the grounds and vicinity of the small but rather pleasant hotel, as well as around the research station (and even along the main road from Mayarí town), with the following species being likely: Stygian Owl, Cuban Green Woodpecker, Crescent-eyed Pewee, Cuban Solitaire, Olive-capped Warbler, Western Spindalis and Cuban Bullfinch. Also worthy of a visit is the Salto del Guayabo, 13km south-south-west of Mayarí, off the main road, with its pine forests and montane serpentine shrub woodland. There are two trails, one to the waterfall overlook, and the other, which is much steeper, descending the valley to the base of the waterfalls. Both are worth hiking. Giant King-bird has been found in the latter region, along with Bee Hummingbird. Grey-headed and Blue-headed Quail-Doves (the latter very scarce) have recently been reported to occur in the vicinity of the Salto, but we are unable to verify this. Anyone choosing to base him or herself in Mayarí may wish to explore the nearby Bahía de Nipe, which is known to be an excellent area for waterbirds and also has Gundlach's Hawk.

## Pico Turquino

Seven kilometres north-west of the village of Ocujal, in Santiago de Cuba province, Pico Turquino is Cuba's highest point, reaching 1,972m. Different habitats found on the peak, which lies within a 17,450ha national park created in 1991, include semi-deciduous forest, evergreen forest, submontane seasonal rainforest, montane tropical rainforest and cloud forest at the summit. From Santiago de Cuba, take the coastal road for about 30km west to the village of Las Cuevas, from where it is necessary to walk along an obvious dirt road to La Demajagua and then follow a trail which leads to the summit. The total walk is about 5 hours (without stops for birding) from Las Cuevas to the lower summit, known as Pico Cuba (1,872m), where the habitat becomes cloud forest, and thence another hour to Pico Turquino. Thus, if you wish to reach the highest areas of the mountain a very early start is essential. It is possible to camp at Pico Cuba or Las Cuevas, but most people will probably prefer one of Santiago's many hotels. While the area is noted for its many endemics (including Cuban Solitaire and Blue-headed Quail-Dove), perhaps the principal avian reward for those willing to climb as high as the cloud forest is the possibility of Bicknell's Thrush. The mountain appears to be a major wintering site in Cuba for this globally threatened species, and possibly the easiest place to find it in the country, although a tape of its winter vocalisation will assist immeasurably in seeing one. Black-capped Petrel perhaps breeds in the region of the summit, but this demands confirmation. Enterprising birders and ornithologists could make a real contribution by spending time in the area searching for this globally threatened seabird.

It is also possible to reach the summit of Pico Turquino via the mountain's north slope, from the scattered settlement of Santo Domingo, which is 18km south of Bartolomé Maso. While it is at least a

12-hour walk from Santo Domingo to the peak, and therefore far less preferable a route, the village houses a noted hotel of the same name, in the grounds of which it is possible to see many of the commoner endemics, such as Cuban Pygmy Owl, Cuban Trogon, Cuban Green Woodpecker, Cuban Vireo and Cuban Grassquit.

## Cabo Cruz

Situated about 36km south of Niquero, the best accommodation is at Marea del Portillo 15km east of Pilón, which has three popular resort-style hotels, in the foothills of Sierra Maestra, about 100km to the east. There is also a reportedly adequate hotel in Niquero and bear in mind that the route from Pilón is somewhat tortuous and long. The area possesses a mixture of coastal thickets and dry evergreen woods, and the Parque Nacional Desembarco del Granma (where Castro, Che Guevara and the other 80 rebels that initiated the Cuban revolution made landfall in the yacht *Granma* from Mexico) has recently been declared a UNESCO World Nature Heritage Site, a national park, and an Important Bird Area. The most significant feature is the cliffs on the south coast, where a small colony of White-tailed Tropicbird breeds, but there are also Cuban Vireo, Oriente Warbler, Cuban Gnatcatcher, Rose-throated Parrot, Fernandina's Flicker, Giant Kingbird and other common endemics in the area's hinterland forests. White-crowned Pigeon can be found at Cabo Cruz itself. From Niquero, drive south through Belic and Las Coloradas to Cabo Cruz.

## Ojito de Agua

This is the area where the Ivory-billed Woodpecker was last definitely found, in 1987. The main habitat type is pine forest. A seven-hour drive from Guantánamo, it is hard to reach. There is no accommodation or locally available food; the nearest town with such facilities is Baracoa, although access is easier from Guantánamo. Permission to visit the area is required and is only issued to *bona fide* researchers, who should apply at least one month in advance from the Agencia del Medio Ambiente (CITMA) office in Guantánamo, directly opposite the Guantánamo Hotel. From the latter, take the main highway east towards San Antonio del Sur, and at the last roundabout on the edge of town go straight on, along a more minor road, towards Jamaica (9km). Thenreafter, continue north through Felicidad (another 20km) and on to Palenque (17km), always following the main road. Beyond Palenque, the route becomes very complex and we recommend that persons interested in visiting the area hire a guide in the latter town. Alternatively, you may enquire at the CITMA office. Given that a broad range of habitats are found in the area, a great diversity of birds occurs here: more than 100 species have been reported, including Plain Pigeon, Grey-headed and Blue-headed Quail-Doves, Cuban Parakeet, Stygian Owl, Bee Hummingbird, Giant Kingbird, Cuban Solitaire and Gundlach's Hawk, making some of the difficulties in accessing Ojito de Agua worth the effort for those who want to get far off the beaten track.

# JAMAICA

The very mountainous island of Jamaica is the third largest in the Caribbean (measuring 230 × 83km), and has a human population of about 3.5 million, of which 95% trace their bloodlines to Africa. English is the official language, although the accent and use of slang make it sound quite different. Tourism is the main industry and most visitors arrive in Montego Bay at the Sir Donald Sangster International Airport. The capital city of Kingston, in the south-east of the island, has a population of over 500,000, and is also the world centre of reggae music. Sugarcane, used in the manufacture of rum, is the main crop. Limestone formations cover over two-thirds of the island, and the highest point is the 2,256m Middle Peak in the 'Grand Ridge' of the Blue Mountains. The forests of the central mountains harbour rushing rivers branching into many small tributaries and waterfalls. Mangroves are evident in coastal areas, but there are also beautiful beaches along the northern coast, where most of the all-inclusive tourist resorts are sited, particularly in Montego Bay, Ocho Rios and Negril.

*Map of Jamaica*

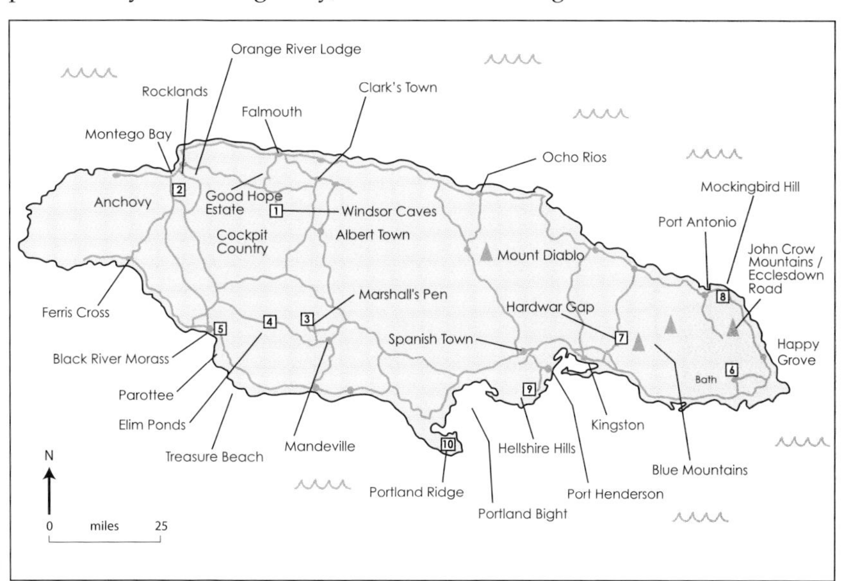


Despite being smaller than either Cuba or Hispaniola, Jamaica sports more endemics than any other of the Greater Antilles, or indeed any other island in the West Indies. It is particularly rich in hummingbirds, tyrant flycatchers and vireos. Thirty endemic species are generally recognised, but two of these, Jamaican Petrel and Jamaican Poorwill, have not been seen since the 19th century and are generally considered to be extinct. Approximately 35 species are restricted to Jamaica or adjacent islands, making it a critically important Endemic Bird Area. Another species, Golden Swallow, which also occurs on Hispaniola, has declined dramatically, there having been no confirmed records on Jamaica since 1989 (in the Blue Mountains, with a 1982 record from Cockpit Country). Birders with sufficient time at their disposal and willing to hire local guides could easily combine a short trip to Jamaica with a longer visit to Cuba, thus giving them chances at a total of over 60 Greater Antillean endemics in just two weeks. There are regular flights in and out of Cuba from both Kingston and Montego Bay. It should be possible to target all of the island's endemics by visiting a maximum of three different sites, and one way to 'tackle' Jamaica is to land at one airport and exit the country at the other, which usually does not attract any additional charge.

*Strategy*

To complete a comprehensive birding trip and maximise your chances of all of the endemics, a trip of five days is ideal. One economical method of achieving this is fly into Montego Bay and purchase a one-way air fare to Kingston; on arriving in Kingston rent a vehicle one-way to Montego Bay. You can drive from Kingston to Bath, through Hardwar Gap, on to Marshall's Pen and back to Montego (to see Black-billed Streamertail you will have to venture east of Kingston to Bath, or east of Port Antonio on the north side of the island). This will save you time and money in the long run. If everything goes well, you could score all of the endemics in as few as three days. If you are on a cruise ship, with only hours for birding, try to arrange a car and driver in advance. From Montego Bay go directly to Windsor Caves. You should be able to see most of the island endemics in this one site in four or five hours. (A six-hour trip to this site in January 1996 produced 21 Jamaican endemics and six multi-island endemics.) If you are on a family holiday to Ocho Rios or Montego Bay, a single day spent at Windsor Caves or Mockingbird Hill will permit a great birding experience without destroying yet another family vacation.

# The North-west

## Windsor Caves *(See Site 1 on Jamaica map)*

If you have only very limited time ashore, for instance if calling into Montego Bay on a cruise, you can easily see 20 Jamaican endemics and six multi-island endemics in the vicinity of this locality at the northern edge of the famous Cockpit Country. If you are able to overnight in the area, it is also one of the best sites on the island for Northern Potoo and Jamaican Owl as well, making it arguably the best single place to bird on Jamaica.

*Location*

Take the A1 from Montego Bay east along the coast as far as Falmouth. Some 0.6km before the Falmouth Inn turn right at the sign for Martha Brae and continue 2.3km to Hague. A road 3.6km beyond here crosses a river, but instead of taking it continue straight ahead to Perth Town, and after 9.1km the road forks – take the right-hand option and continue until you reach a T-junction, after 14.4km, at Sherwood Content. Turn right and continue to a left fork at 14.8km. From here proceed to another T-junction where you turn left (19.9km). The road terminates 20.4km from Falmouth at another T-junction. Park nearby and hike uphill from the shack by the water tower with the bird graffiti.

*Strategy*

There are usually several flowering trees near where you will have parked, and Jamaican Mango and Red-billed Streamertail are both numerous in this area. Pass through a field, which borders the forest; Jamaican Vireo, Jamaican Pewee and Stolid Flycatcher can all be found here. Upon entering the forest the trail splits. The short left fork ends at a cave mouth, which area is good for Blue Mountain Vireo, but comparatively little else. The right track continues through forest. On arriving at a large semi-cleared area on the right-hand side, check for both Chestnut-bellied and Jamaican Lizard Cuckoos. Many other endemics can be found along this section of trail, which eventually reaches a cultivated area, then continues beyond a small farmhouse. There should be little need to go much further. Both Jamaican Owl and Northern Potoo are best searched for in the vicinity of Windsor House.

***Accommodation*** While you could day-trip this site from Montego Bay, using any of the great many hotels in the town as a base (there are several relatively less expensive options in close proximity to the international airport), it is also possible to stay at the nearby Windsor Great House. This will also give you more time and chances for night birding. If you want to spend the night at the Great House it is best make reservations before you arrive in Jamaica. Contact Michael Schwartz, Windsor Research Station, Windsor Great House, Sherwood Content PO, Trelawny, Jamaica (telephone 876-997-3832/fax 876-954-7716/e-mail: windsor@cwjamaica.com or michaelschwartz@cockpitcountry.com or visit www.cockpitcountry.com/bedrooms.htm to book in advance. It is important to leave Montego Bay while there is daylight, as the forks and turns in the road can be easily missed in the dark.

***Birds*** The following single-island endemics can all be found in the vicinity of Windsor Caves: Ring-tailed Pigeon, Crested Quail-Dove, Chestnut-bellied and Jamaican Lizard Cuckoos, Jamaican Mango, Red-billed Streamertail, Olive-throated Parakeet (introduced in Dominican Republic), Yellow-billed and Black-billed Parrots (both parrots regularly overfly the area, but are difficult to see perched), Jamaican Owl, Jamaican Tody, Jamaican Woodpecker, Jamaican Pewee, Sad Flycatcher, Jamaican Becard, Jamaican Crow, Blue Mountain and Jamaican Vireos, Arrowhead Warbler, Jamaican Spindalis, Orangequit, Yellow-shouldered Grassquit and Jamaican Oriole. Other species of interest, including multi-island endemics, include: White-crowned Pigeon, Zenaida and White-winged Doves, Ruddy Quail-Dove, Vervain Hummingbird, Antillean Palm Swift, Northern Potoo (drive nearby roads checking fence posts) Loggerhead Kingbird, Stolid Flycatcher, Greater Antillean Grackle, Yellow-faced and Black-faced Grassquits, and Greater Antillean Bullfinch.

## Rocklands Feeding Station *(See Site 2 on Jamaica map)*

This famous site forms a well-known part of the tourist circuit and can be very busy with 'regular' tourists on some days. Hardcore birders will probably mainly wish to visit the feeding station as something to do on their last day in the country, perhaps before getting a flight home. Nonetheless, its very handy location close to Montego Bay cannot be ignored, especially perhaps if you are on a non-birding holiday with a spouse who does not share your interests. One can enjoy a reasonable range of endemics and migrant warblers in season, as well of course as the spectacle of hummingbirds that will perch on your hand.

***Location and Strategy***

Take the A1 south-west from Montego Bay to Reading. If you leave in the early morning, it would be worth briefly continuing on the A1 beyond Reading as far as the Tryall Golf & Beach Club, where a search of the golf course ponds early or late in the day can be profitable for Yellow-breasted Crakes, which forage along the shorelines.

To get to Rocklands take the B8 at Reading south towards Anchovy, and 3.5km from this junction there is an unpaved road on the right signposted for Rocklands Feeding Station. After 0.2km turn right up the hill and 0.9km from the Anchovy road turn-off you will see the entrance to the feeding station on the right. Caribbean Doves are seen in this area, so don't hurry.

Lisa Salmon, the well-known owner of the feeding station, died in the year 2000, but her caretaker Fritz has taken over in her stead. He is a great guide, if somewhat expensive. A smaller fee is payable if just visiting the feeding station, which is only operational between 15.00 and 17.00 daily. There are a couple of short trails that enter the forest behind the Rocklands buildings, and these are definitely worth walking if you have time. To contact the feeding station in advance of a visit, telephone 809-952-2009.

***Accommodation***

Most, if not all, visitors will stay in Montego Bay. If you arrive in Montego Bay in the evening, you may prefer to wait until morning to rent a car. If you wish to stay in Montego a good spot is the Relax Resort (telephone: 876-979-0656 / fax: 876-952-7218 / e-mail: relax.resort@cwjamaica.com) at about $100 / night for a double, or $60 for single. See also the section on Windsor Caves. If you wish to get out of the resort area, the similarly priced Fisherman's Inn lies 33km east of Montego Bay, at Falmouth (telephone: 876-954-3427 / fax: 876-954-4078), which is near the turn-off for Windsor Caves. Others have also recommended Emerald View Villas, a locally owned guesthouse about 20 minutes east of Montego Bay, with dry woodland close by (http://emeraldview.net/).

***Birds***

Single-island endemics (most of which must be searched for along the trails behind the feeding station) regularly seen here include: Olive-throated Parakeet, Yellow-billed Parrot, Jamaican Lizard, Chestnut-bellied and Mangrove Cuckoos, Jamaican Mango, Red-billed Streamertail, Jamaican Owl (infrequent), Jamaican Woodpecker, Jamaican Elaenia, Rufous-tailed and Sad Flycatchers, Jamaican Becard, White-eyed and White-chinned Thrushes, Jamaican Vireo, Jamaican Crow, Arrowhead Warbler, Jamaican Spindalis, Jamaican Oriole, Orangequit and Yellow-shouldered Grassquit, while Jamaican Blackbird is recorded occasionally. Other species of interest, including multi-island endemics, include: Caribbean Dove (one of the best sites on the island), Ruddy Quail-Dove, Northern Potoo, Stolid Flycatcher, Loggerhead

Kingbird, Black-whiskered Vireo (summer), Greater Antillean Grackle, Saffron Finch, Black-faced and Yellow-faced Grassquits, and Greater Antillean Bullfinch.

Arrowhead Warbler

**OTHER SITES**

If you continue along the Anchovy road south, turn off the B8 at Shettlewood and look for a small pond on the left; Masked Duck can be found here. At **Shettlewood** the B7 branches off southward; to go to the Black River, turn off the B7 at Baptist south on the A2 and continue to the Black River. Alternatively, to get to Marshall's Pen, go in the opposite direction (north and east) from Baptist along the Bamboo Avenue (as this stretch of the A2 is known) to Mandeville. (See Jamaica map.)

To visit the **Orange River Ranch** in Montego Bay take Queen's Drive east to the traffic circle, then St. James (the upper road) on the left and continue to Barnett. The last block before Barnett is one way in the wrong direction, so turn left a block after Barnett, go to the end and turn right then left onto Barnett. Go to Fairfield (there is a McDonalds and Westgate shopping centre is on the left), which is marked, and you will see a sign for the ranch; turn left and follow the signs, some of which are obscured by foliage. Just before Johns Hall, you will see the ranch clearly marked on the left. This property of roughly 400ha, which has accommodation (e-mail: WhitterGroup@mail.infochan.com / website: www.montego-bay-jamaica.com / orangeriver / telephones: 876-601-3987, 876-919-1017 or 876-979-3294), also makes for a convenient short-duration site out of Montego Bay. The ranch offers a complementary shuttle service to and from Montego Bay and the international airport, for guests. Species of interest here include: Caribbean Dove, Yellow-billed Parrot, Red-billed Streamertail, Vervain Hummingbird, Rufous-tailed Flycatcher, Jamaican Vireo, Black-whiskered Vireo (summer), Jamaican Crow, Jamaican Spindalis, Jamaican Euphonia, Orangequit and Jamaican Oriole.

If travelling from Montego Bay to Mandeville, the road from Jackson Town to Christiana, via Ulster Spring, Albert Town and the evocatively named Wait-A-Bit, can be very productive throughout its length for the **Cockpit Country** specialities, namely Jamaican Crow and Black-billed and Yellow-billed Parrots. All three species should be spotted from the road, if you keep a sharp lookout, and stops might also produce less common endemics such as Jamaican Euphonia. The other road across this part of the Cockpit Country, from Kinloss to Albert Town, via Barbecue Bottom and Burnt Hill (although marked as a town on most maps, there are no buildings there), was in very rough condition a few years ago and is probably best avoided, although both parrots and the crow, as well as Ring-tailed Pigeon, Caribbean Dove, Blue Mountain Vireo, Arrowhead Warbler and White-chinned Thrush, can all be found along this route. Around St. Vincent, just west of Albert Town, the Jamaican endemic Ring-tailed Pigeon can also be found.

Another site for the Cockpit Country specialties is the **Good Hope Estate**, which is reached by heading inland on the road from Falmouth to Martha Brae, then continuing towards Bounty Hall, but turning off left on a minor road signed to Good Hope. Just before the town of Good Hope itself, the estate (accommodation available) is signed off to the right. Both *Amazona* parrots, Jamaican Crow and Jamaican Becard can all be found here, sometimes from the road itself (park carefully).

# The Central-South

## Marshall's Pen, Mandeville *(See Site 3 on Jamaica map)*

Marshall's Pen, a 300ha cattle property and nature reserve, is located on the outskirts of Mandeville. It is the home of Ann Sutton, whose husband Robert, the island's pre-eminent ornithologist, was tragically murdered in July 2002. Ann, an ornithologist in her own right, and her guide Brandon Hay, are currently carrying on the tradition of providing a fine base for birders visiting Jamaica.

***Location***

To reach the pen from the west, follow the signs to Mandeville and turn left onto the first bypass at the first roundabout as you enter the town. After 1.4km, turn left towards Somerset Quarries, and after 1.5km turn right at the T-junction. Follow the road, which winds left then right, and at 2.1km from the roundabout you will see the entrance to Marshall's Pen on the right-hand side of the road between two stone pillars. After 2.7km go right through the gate, and continue to the main house. To return to Negril just take the A2 westward (64km). To return to Montego Bay there are many options (see the map). To go to Kingston take the A2 east (120km) – see Jamaica map.

*Map of Marshall's Pen*

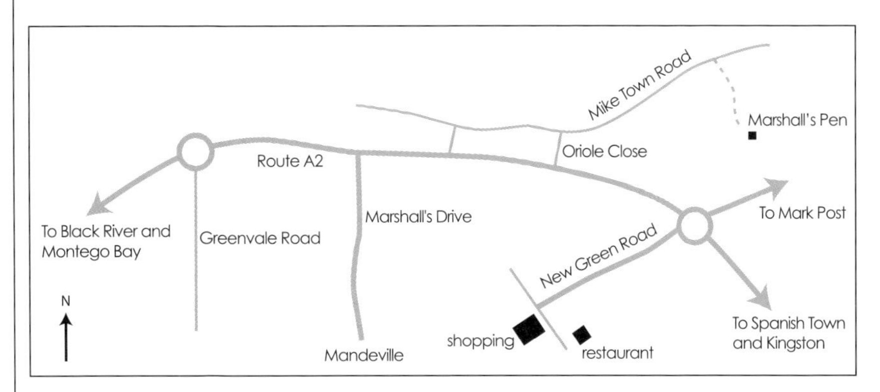


*Accommodation*

Ann Sutton (e-mail: asutton@cwjamaica.com) offers accommodation on either a full-board, bed-and-breakfast or self-catering basis. If choosing the latter, it is just a short drive into Mandeville, where there is a range of eating possibilities. It is extremely wise to book in advance, especially if your group is a large one (Ann can cater for groups in excess of ten people) or if visiting during the winter or early spring.

*Strategy*

Most birds can be found within close proximity of the main house; a series of trails, which can be very muddy, access a mosiac of fields, woodlots and borders. Ann and her guides will usually know the whereabouts of most of the species. Jamaican Becard often nests in the area, and Jamaican Owl and Northern Potoo are frequently found in the very close environs of the house (the latter sometimes at its day roost). Crested Quail-Dove (as well as other doves) and White-eyed Thrush are frequently found by walking the trails as quietly as possible. Hummingbird feeders ensure good views of the streamertail and mango.

*Birds*

This is arguably the best site on Jamaica for Jamaican Owl and White-eyed Thrush. Other single-island endemics include: Crested Quail-Dove, Olive-throated Parakeet, Chestnut-bellied and Jamaican Lizard Cuckoos, Jamaican Mango, Red-billed Streamertail, Jamaican Tody, Jamaican Woodpecker, Jamaican Pewee, Jamaican Elaenia, Rufous-tailed and Sad Flycatchers, Jamaican Becard, White-chinned Thrush, Jamaican Vireo, Jamaican Crow, Arrowhead Warbler, Jamaican Euphonia, Jamaican Spindalis, Jamaican Oriole, and Yellow-shouldered Grassquit. Other species of interest, including multi-island endemics, include: Least Grebe, White-crowned Pigeon, Zenaida, White-winged and Caribbean Doves, Ruddy Quail-Dove, Vervain Hummingbird, Black and Antillean Palm Swifts, Northern Potoo, Antillean Nighthawk (summer), Stolid Flycatcher, Grey Kingbird, Loggerhead Kingbird, Rufous-throated Solitaire, Black-whiskered Vireo (summer), Cave Swallow, Yellow-faced and Black-faced Grassquits, Saffron Finch, Greater Antillean Grackle and Greater Antillean Bullfinch. In the boreal winter the area can be very productive for warblers and other passerines that breed in North America, and a local specialty is the introduced European Starling!

**OTHER SITES**

**Elim Pools** *(See Site 4 on Jamaica map)* is a great site for waterbirds, especially the globally threatened West Indian Whistling Duck (up to 75 reported in recent years) and Caribbean Coot. A variety of herons (including Least Bittern), ibises, Limpkin, and Least and Pied-billed Grebes are also commonly seen. Several rails are possible, including Spotted and Black, but only Sora is especially likely to be encountered. Particularly during passage periods the area can also be productive for many shorebirds, and a few endemics, such as Jamaican Woodpecker and Jamaican Euphonia, can also be found. Mangrove Cuckoo is also possible here, and Antillean Nighthawk in summer. Vagrants sometimes appear here; there was, for instance, a Black-bellied Whistling Duck seen here in February 2008. The pools, which can be very busy with anglers at weekends, are located off the A2 between Mandeville and Middle Quarters. At the town of East Lacovia, there is a Texaco garage on the north side of the road. From here, look for a road signed to Maggotty. Turn here, zero your odometer, and drive through Haughton. After 6.6km, at the village of Newton, turn right at the fork, and cross the stream. At 6.8km turn right along this stream onto a dirt road. At 13.4km another track takes off to the left (east) at a large pumping station. Open water soon appears on the right. Dusk is the best time for whistling ducks, which are perhaps most frequently seen in flight. To return to Mandeville go back to the A2 and proceed east.

**Black River Morass** *(See Site 5 on Jamaica map)*, a recently declared Important Bird Area and Ramsar site, encompasses 5,700ha of mangrove, lagoons, swamps, forests and estuaries. This is another good site for West Indian Whistling Duck, which is best seen by taking the two-hour boat tour, which originates at the Black River Bridge. Piping Plover has been recorded on the beaches here, and the area is also important for the numbers of White-crowned Pigeon present. Black River Safari Boat Tours (www.silver-sands.com/black_river_safari.html) cater to birders. The night cruise is the best for the whistling duck, while a private tour to the Middle Quarters River offers the best chance of Yellow-breasted Crake.

Around **Treasure Beach**, on the south coast, there are various ponds that can be productive for the localised Masked Duck, although numbers apparently fluctuate substantially with water levels. One of the consistently best freshwater ponds is that at Treasure Beach itself, which can hold Caribbean Coot, as well as ducks, waders, herons and other waterbirds, but to find Masked Duck it may be necessary to explore the area as far north as Hill Top and Parottee, where the ponds can be very productive for shorebirds in winter. The brackish waterbody, Great Pedro Pond, just east of Treaure Beach, is another important area to check. Ann Sutton owns a self-catering cottage here, which can be booked by contacting her in advance (see Mandeville section).

# The East

## Bath *(See Site 6 on Jamaica map)*

Kingston does not have the reputation of being a very friendly city, but there is some reasonable birding close by. One possibility for Black-billed Streamertail is the Bath Fountain Hotel. This hummingbird is sometimes also found on the east side of the town of Morant Bay, but Bath represents a surer bet.

***Location*** From Kingston, take the A4 main coast road east to Morant Bay and continue almost as far as Port Morant, where a more minor road cuts inland (left) through Harbour Head and Ginger Hall to Bath. The Bath Fountain Hotel is signed left off the main road through the small town of Bath.

***Accommodation*** The hotel and on-site restaurant at Bath (telephone: 876-703-4345) itself makes a perfect base, if you decide to stay the night in the area. Another good place to overnight is Morant Bay Villas (telephone: 876-982-2418) at Morant Bay.

***Strategy*** The main, if not the only, reason for visiting is to find Black-billed Streamertail, for which Bath Springs is probably the closest site to Kingston, should this be your entry or exit point to the country. The hummingbird is easily found at the feeders.

***Birds*** The following single-island endemics can be found in the vicinity of the hotel: Black-billed Streamertail, Jamaican Tody, Rufous-tailed and Sad Flycatchers, White-chinned Thrush, Jamaican Euphonia, Jamaican Spindalis and Orangequit. Other species of interest, including multi-island endemics, include: White-crowned Pigeon, Antillean Palm Swift, Loggerhead Kingbird, Yellow-faced Grassquit, Greater Antillean Bullfinch and Greater Antillean Grackle.

## Hardwar Gap and Hollywell National Park

*(See Site 6 on Jamaica map)*

If approaching the area from downtown Kingston, follow the signs to Halfway and New Kingdom, then signs to Papine. Take Hope Road, then Old Hope Road, which joins Gordon Town Road. Follow the signs to Gordon Town, but before you reach Gordon Town you will see a sign for the Gap on the left. Follow this road, the B1, from Newcastle Square via Irish Town to the Gap. It is also possible to approach the road from the north, from Buff Bay. If you decide to visit Bath to find Black-billed Streamertail prior to the Gap, then it is best to return to Kingston along the coast road, rather than attempting to avoid the capital by following any minor roads.

***Strategy*** Low cloud and rain are frequent features of Hardwar Gap. The birding can be excellent and a great many of the endemics can be found in the area, but bear in mind that it is easily possible to lose quite some time to poor weather. Approximately 3km south of the Gap, on the left-hand side when travelling north, you will see Woodside Drive, which slopes steeply downhill. Park carefully along the main road and walk

*Map of Hardwar Gap*

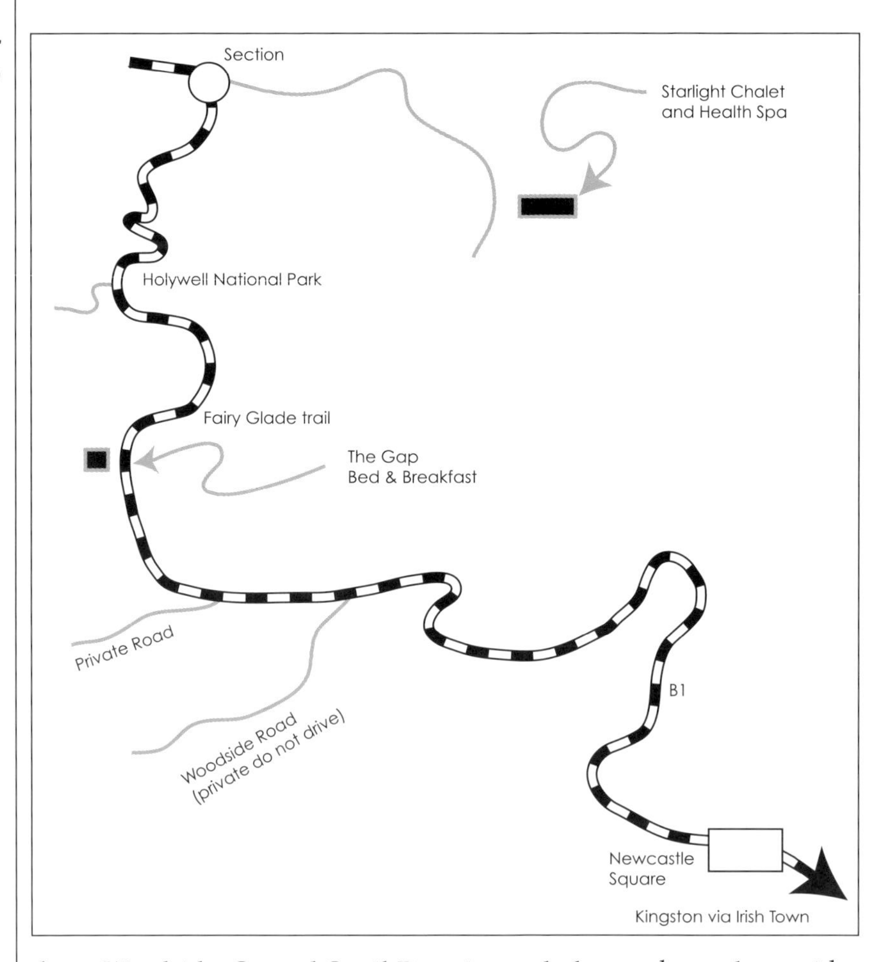


down Woodside. Crested Quail-Dove is regularly seen here, along with Ring-tailed Pigeon and several other endemics. Just over 1km from the road, there is a reasonable overlook, which affords the best chance of the pigeon.

The quail-dove is also frequently seen along the main road at quieter times of day with less traffic. Indeed, the entire B1 is excellent for birding from Irish Town to Section, but that part between the Gap restaurant and just beyond Hollywell Park is probably consistently the most productive. The Fairy Glade trail, which starts a few metres beyond the Gap on the opposite of the road to the restaurant, has been productive in the past, but is frequently rather overgrown and is probably not to be favoured over the road itself. The best sites for the Jamaican Blackbird are 1.3km north of the Hollywell Park entrance, where the road makes a hairpin turn to the left. There is a steep ravine here full of bromeliads, where the birds frequently feed. They can be extremely elusive and unobtrusive, but sometimes respond well to playback, although still approaching the observer quietly. Alternatively, there are many trails in the park (making a guide potentially useful), of which the Waterfall Trail is arguably the best. Crested Quail-Dove, Rufous-throated Solitaire, Blue Mountain Vireo, White-eyed Thrush and Jamaican Blackbird can all be seen along here.

**Accommodation**

There is a restaurant at the Gap, which is generally open from 10.00 to 17.00 on weekdays, and until 18.00 at weekends. Reservations are definitely required for supper (telephone: 809-997-3032). The owners sometimes have self-catering accommodation available. The feeders here can be very busy with Red-billed Streamertails (best watched while enjoying the expensive, but fabulous and deservedly world-renowned 'Blue Mountain Coffee'). The Greenhills Guesthouse nearby makes a good place for lodging. Other places to stay near Hardwar Gap, all of which provide food, include Strawberry Hill, at Irish Town (telephone: 808-944-8400; website: www.seejamaicacheaply.com/hotels/kingston_jamaica/strawberry_hill.html); Blue Mountain Inn, on Gordon Town Road (telephone: 809-927-1700); Forres Park, Mavis Hill, near St. Andrew (www.neilmar.com/forres/forres_test.php); and the Starlight Chalet and Health Spa, in St. Andrew (telephone: 876-969-3070/924-3075). There are also cabins at Hollywell Park, which should preferably be booked in advance via the Jamaican Conservation and Development Trust (JCDT; e-mail: jcdt@kasnet.com), although it may be possible to just turn up.

**Birds**

One the best sites on the island for a wide range of Jamaican endemics including: Crested Quail-Dove, Chestnut-bellied and Jamaican Lizard Cuckoos, Red-billed Streamertail, Jamaican Tody, Jamaican Woodpecker, Jamaican Pewee, Jamaican Elaenia, Rufous-tailed and Sad Flycatchers, Jamaican Becard, White-eyed and White-chinned Thrushes, Jamaican and Blue Mountain Vireos, Jamaican Crow, Arrowhead Warbler, Jamaican Euphonia, Jamaican Spindalis, Orangequit, Jamaican Blackbird, and Yellow-shouldered Grassquit. Other species of interest, including multi-island endemics, include: White-crowned Pigeon, Ruddy Quail-Dove, Vervain Hummingbird, White-collared Swift, Greater Antillean Elaenia (one of the few regular sites on the island for this species, which is otherwise restricted to Hispaniola), Stolid Flycatcher, Loggerhead Kingbird, Rufous-throated Solitaire, Black-whiskered Vireo (summer), Yellow-faced and Black-faced Grassquits, Greater Antillean Grackle and Greater Antillean Bullfinch. In the boreal winter the area can be very productive for warblers and other passerines that breed in North America, amongst them the secretive Swainson's Warbler.

## Mockingbird Hill *(See Site 8 on Jamaica map)*

This very comfortable hotel (which borders on the luxurious) is located just outside Port Antonio, in the extreme north-east of the island, and is perhaps the perfect place to bring a non-birding spouse for a relaxing vacation, which also permits you a good chance of seeing all of Jamaica's endemics. Those hell-bent on seeing all of the island's specialities, but with limited time at their disposal, might also choose to 'bite the bullet' and spend a bit of extra money to maximize their time and chances. It is possible to find all of the strict Jamaican endemics in just two days in this area, if visiting the nearby Ecclesdown Road area. There are many other hotels in Port Antonio suiting a wider range of budgets, but the Mockingbird Hill folk can help you to organise a bird guide if you need one.

***Location*** If accessing the site from Kingston (two and half to three hours drive), continue north on the B1 beyond Hardwar Gap, to Buff Bay, and then turn right and continue east on the A4. Alternatively, you can follow the coast road, first east towards Bath and then north and west. Black-billed Streamertail can be found at several sites between here and Port Antonio, but the best site is Mockingbird Hill, which is also an extremely productive area for a great many of Jamaica's single-island endemics. The hotel is just 7km outside of Port Antonio, inland of the A4 to the east of town and is well signed. Indeed, if you land in Montego Bay and want to visit just one area on Jamaica, but do have time to get over to the east end of the island (the drive takes at least 4.5 hours), then it might be best to focus on Mockingbird Hill, although you might still need to make some day trips, e.g. towards the Hardwar Gap region, to have a chance of seeing all of Jamaica's endemics. However, do bear in mind the distance on the slow coast road, initially the A1 between Montego Bay and St. Ann's Bay, which becomes the A3 just west of Ocho Rios and then the A4 between Annotto Bay and Buff Bay.

***Accommodation*** Mockingbird Hill is becoming an increasingly popular (but quite expensive) option for visiting birders, so it is almost certainly advisable to book in advance, especially if you want to book any of their more budget-priced accommodation (the hotel has a range, and can provide meals, either as part of a pre-paid plan or a la carte on site). They can also arrange birdwatching guides, booked through the Jamaica Conservation and Development Trust, and transport to and from either Kingston or Montego Bay, although these transfers are not cheap. E-mail info@hotelmockingbirdhill.com, visit the hotel's web pages at www.hotelmockingbirdhill.com, or telephone 876-993-7267.

***Strategy*** Many of the island's endemics can be found on the hotel's property, but the trail at Sherwood Forest, which is accessed by continuing east along the main A4 coast road, then turning inland just before the village of Fairy Hill, should be explored if you have no luck elsewhere with Jamaican Blackbird. If booking a guide through the hotel (as of early 2009, two of the best were Roger and Ryan), he or she should be able to advise you on local sites in the John Crow Mountains or Blue Mountains for any other needed endemics.

***Birds*** The following single-island endemics can be seen in the grounds of the hotel: Ring-tailed Pigeon (especially in the boreal spring), Olive-throated Parakeet, Chestnut-bellied and Jamaican Lizard Cuckoos, Jamaican Owl (sometimes in the garden but more regularly found by the houses at the foot of the hill by the A4 coast road; the staff should have up-to-date information), Jamaican Mango, Black-billed Streamertail, Red-billed Streamertail (heavily outnumbered by the previous species, but reasonably regular), Jamaican Tody, Jamaican Woodpecker, Jamaican Elaenia, Sad and Rufous-tailed Flycatchers,

Jamaican Oriole

Jamaican Becard, Jamaican Vireo, Jamaican Crow, White-chinned Thrush, Jamaican Euphonia, Jamaican Spindalis, Orangequit and Jamaican Oriole. Other species of interest present in the general vicinity of the property, including multi-island endemics, include: White-crowned Pigeon, Caribbean and Zenaida Doves, Ruddy Quail-Dove, Green-rumped Parrotlet, Northern Potoo, Vervain Hummingbird, Stolid Flycatcher, Grey and Loggerhead Kingbirds, Greater Antillean Bullfinch, Black-faced Grassquit and Greater Antillean Grackle. The nearby San San Bird Sanctuary and Nature Reserve, which is well signed (inland) off the main A4 coast road, less than 2km east of the turn to Mockingbird Hill, should hold many of the same species, although we have not birded this area personally.

**OTHER SITES**

The cliffs at **Happy Grove** can be reached either by driving along the coast road east and south from Port Antonio, or by continuing east from Bath until you reach Hardley and the A4 coast road. Then turn north (left) and continue beyond Hectors River to Happy Grove. The site is only really worth visiting between January (exceptionally November) and April, during which period White-tailed Tropicbird can be found breeding in the area in small numbers. Mornings are perhaps best.

If staying at Mockingbird Hill and or visiting Happy Grove, then it will be well worth following the minor road inland from Long Bay, on the A4 between Port Antonio and Hectors River. Follow the road through the village of Hartford and after 1.1km turn left into the small town of Windsor Forest, before heading more or less south, by the second turn to the left after leaving the coast road, to **Ecclesdown**. The turn is not well signed, but is located next to a shop with blue doors. This site is about 40 minutes drive from Mockingbird Hill. Productive habitat for a rich variety of endemics, including the globally threatened Ring-tailed Pigeon, is reached after a further 3km. Park carefully along this narrow road and search wooded areas for Ruddy Quail-Dove, as well as Crested Quail-Dove (which is most easily found soon after dawn) the introduced Green-rumped Parrotlet, both parrots (Black-billed seems particularly common, but both are numerous), both endemic cuckoos, Jamaican Mango, Black-billed Streamertail, Vervain Hummingbird, Jamaican Elaenia, Rufous-tailed Flycatcher, Jamaican Becard, Blue Mountain, Black-whiskered and Jamaican Vireos, Jamaican Crow, Rufous-throated Solitaire, both endemic thrushes, Arrowhead Warbler, Yellow-shouldered Grassquit, Orangequit, and Jamaican Blackbird, which seems very easy to find in this area, among many other species. It is possible to loop back to the coast road by continuing beyond Ecclesdown, although the road is not good.

Just outside Kingston, at the north end of the low-lying **Hellshire Hills** *(See Site 9 on Jamaica map)*, the dry forest around the old **Port Henderson** ruins provides the chance of a number of endemics. Follow Marcus Garvey Drive west along the seafront through the capital. Immediately beyond the far end of Tinson Pen local airport, which is on the north (right) side of Marcus Garvey Drive, turn off left along the causeway road, which crosses Hunts Bay. Continuing along South Causeway, turn left, just before the road swings around to the right and merges into Dawkins Drive, on Augusta Drive. On reaching Port Henderson take the left-hand exit at the first roundabout; this road leads to the ruins.

Those with more time to explore this habitat type would be better advised to head for the **Portland Ridge** *(See Site 10 on Jamaica map)*, part of the 87,615ha Portland Bight Protected Area, south of May Pen, on the A2 between Kingston and Mandeville. From May Pen, head south on the B12 towards Hayes and Lionel Town. Beyond the latter the road is sometimes in very poor condition, but keep heading south towards Portland Cottage, where you should turn left along the minor road that crosses the Portland Peninsula to a lighthouse. Stop and explore any suitable areas of dry limestone forest (all of which is secondary), especially in the first few kilometres after Portland Cottage, searching for Plain and White-crowned Pigeons (which are hunted between August and September), Caribbean and White-winged Doves, Jamaican Mango, Jamaican Woodpecker, Jamaican Vireo, Bahama Mockingbird, and Jamaican Oriole. West Indian Whistling Duck is present in this area, although the mockingbird is arguably the key target at both these last two sites, but is far more easily found on Cuba. If you are planning a two-island holiday, including Cuba, it would be certainly more advisable to search for this species on the latter island, as a number of key endemics share the same area as that where Bahama Mockingbird occurs.

# HAITI

The landscape of Haiti is demarcated sharply from the Dominican Republic, a fact readily observed when flying over of the island of Hispaniola, or when standing on the border between the two countries. While Haiti appears barren and brown, in comparison the Dominican Republic is lush and green. A depressing 98% of Haiti's original forest cover has been destroyed, a by-product of which has been the destruction of much fertile agricultural land. Subsistence agriculture is nevertheless practiced. Over half of the population of over 8.5 million is illiterate and a staggering 80% (or more) live in poverty. Foreign aid comprises 30–40% of the government budget, and Haiti has the sad distinction of being the poorest country in the Western Hemisphere. It is overpopulated and has suffered from many years of political upheaval and economic decline all of which have been spectacularly exacerbated by the 2010 earthquake which claimed over 200,000 lives. Despite the current presence of UN peacekeeping troops, a visit to the country is not to be considered lightly, as kidnapping has recently become commonplace and some women victims have been raped.

On arriving at Port-au-Prince international airport, you would be well advised to find a rental car and leave the capital city immediately. The streets are lined with beggars, and this is definitely not a tourist area! The main objective for any birder to visit Haiti is the Grey-crowned Palm Tanager, which is basically endemic to the country, although there have been just over a handful of reports from the extreme south-western border region of the Dominican Republic, mainly on the road between Perdanales and El Aguacate. However, at least some of these sightings might well have involved immature Black-crowned Palm Tanagers or hybrids between the two species (none seems to have been documented with photos).

*Map of Haiti*

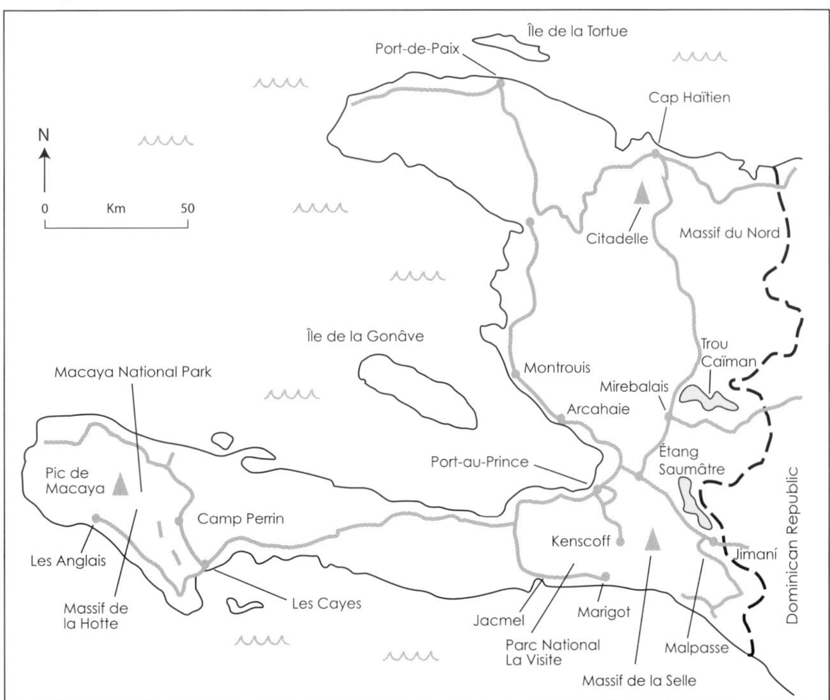

There are three land border crossings between Haiti and the Dominican Republic, but because it is not possible to take rental vehicles from one country to the other, their usefulness is rather limited for visiting birders. The southern crossing point, at Malpasse/Jimaní, is served by public transport to and from Port-au-Prince, and would be most convenient for any very enterprising birders who wished to 'make the run' into Haiti.

## Jacmel

If one is in pursuit of Grey-crowned Palm Tanager, which is the only reason most birders will even consider visiting Haiti, the species will not be observed until you get close to Jacmel, some 80km west of Port-au-Prince, which area was devastated by the recent earthquake. Until this point, Black-crowned Palm Tanager is common, and Grey-crowned noticeable only for its absence. From the international airport proceed south, then west, on Boulevard Jean-Jacques Dessalines (Route Nationale 2) to Dufort, where it forks and from here you should follow the Vers Port-au-Prince (Route 204), which eventually turns south to Jacmel. From Dufort onwards to Jacmel, on the Vers Port-au-Prince, virtually any area of vegetation should produce Grey-crowned Palm Tanager; although it is more common in and around gardens in Jacmel, and further west on the Tiburón Peninsula. Note that the two species apparently hybridise near the Jacmel-Fauché depression.

## Parc Nationale La Visite

This park lies just 22km south of Port-au-Prince in the Massif de la Selle. It boasts nearly 3,000ha of pine forest, savanna and montane cloud forest, at altitudes from 1,900 to just below 2,300m. Any intending visitors are well advised to consult the paper in *Cotinga* 16 (by Liliana Dávalos and Thomas Brooks, published in 2001), which lists observations and further access details based on a visit in January 2000. Those prepared to hike (and perhaps even camp) will likely enjoy the greatest success at this site, as the roads are extremely rugged and in poor shape, and the only accommodation is generally some distance from the highest areas with most birds.

***Location*** Kenscoff, which is one base for exploring the park, is reached from the airport by taking Route de Delmas to the suburb of Petionville. Turn left on Gregaire, which ends at Oge. Turn left one block to Route de Kenscoff and turn left again to Kenscoff, which is reached via the settlement of Fermathe. From Kenscoff you will probably need to hike into the park in the direction of Sequin, which takes up to about six hours, unless you have a four-wheel-drive vehicle (in which case see below). Alternatively, and perhaps preferably, you can drive east from Jacmel on Avenue Baranquia (Vers Cayes-Jacmel) to Marigot and then north into the mountains. From Fond Jean Noel, there is another steep climb to Sequin, and from there is also the major hiking trail from Sequin to Kenscoff.

***Accommodation*** The Hotel La Florville in Kenscoff is a good place to stay (telephone:

509-255-7139). Just beyond Sequin, on the left-hand side of the road, there is a bed-and-breakfast establishment, the Auberge du Visite, which is fairly expensive but can provide guides. It is possible to camp in the park.

***Strategy*** En route to Kenscoff from Petionville, it is worth stopping at the souvenir shop in Fermathe. Take the track through the gate on the right side of the Baptist Mission church; when you reach a grassy playground on the left, you will find an opening to a small woodlot on a steep slope, below which is an area of scrub. Hummingbirds and Green-tailed Ground Tanager, along with some other commoner birds can be found in this area. Migrant warblers are a feature, and Black-whiskered Vireo is present in summer. If it proves possible to drive on above Kenscoff, stay on the paved road left past the police station, then take the right-hand way at the next fork and continue towards the settlement of Furcy. The road is even worse after the fork, but eventually it reaches more level ground at the orphanage known as Saint Helene. From here the main, unpaved, road leads downslope to Furcy through mixed woodland, whilst a narrower road to the left leads up to some communication towers, around which Golden Swallow can be found.

***Birds*** Much of the following information is taken from the paper by Dávalos and Brooks. Hispaniolan and Greater Antillean regional endemics observed in the park include La Selle Thrush (easy to see in the early morning in January 2000), Red-legged Thrush, Rufous-throated Solitaire (best looked for early and late in the day), Hispaniolan Crossbill, Antillean Siskin (both the last two species are best looked for at Pic La Visite), Golden Swallow (not uncommon in January 2000), Narrow-billed Tody (at higher elevations), Broad-billed Tody (lower), Hispaniolan Highland Tanager (perhaps no recent records), Green-tailed Ground Tanager (mainly in moist forest, but seemingly rare), Hispaniolan Lizard Cuckoo (apparently rare), Hispaniolan Trogon, Western Chat-Tanager (best looked for in broadleaf woodland in ravines), Ashy-faced Owl, Vervain Hummingbird, Hispaniolan Emerald, Hispaniolan Woodpecker, Hispaniolan Pewee, Black-crowned Palm Tanager, White-necked Crow (only recorded near Kenscoff, in January 2000), Hispaniolan Palm Crow (widespread throughout the massif), Hispaniolan Spindalis, Antillean Palm Swift (sometimes consorts with Golden Swallows), Palmchat (common below 1,800m), Greater Antillean Bullfinch, Greater Antillean Grackle, Hispaniolan Oriole, Scaly-naped Pigeon, Greater Antillean Elaenia, and Antillean Euphonia (generally quite common). Pine Warbler can be found in suitable areas along the road between Fermathe and Kenscoff, as well as in pine woodland higher in the park. Yellow-faced and Black-faced Grassquits can also be found below Kenscoff, considerable numbers of North American migrants (including many warblers and the globally threatened Bicknell's Thrush) winter in the park and its environs, and there are at least two breeding colonies of the globally threatened Black-capped Petrel known from the park's north-facing limestone scarp, although your chances of stumbling across this seabird are practically none. For those who are intent on finding it, the species is a winter breeder.

# DOMINICAN REPUBLIC

Well known for its cigars, rum and baseball players, the Dominican Republic forms one half of the island of Hispaniola, the second largest of the Antilles and Christopher Columbus' first landfall in the New World. Among modern-day birders, able to cross the Atlantic rather more easily, the country is renowned for its exciting range of endemic birds. Despite this, and the availability of a new first-rate field guide, to some extent the Dominican Republic is still overlooked as a birding destination. Compared to Cuba it lacks a relatively well-developed local birding infrastructure, coupled of course with the aura of being a last bastion of communism and the spiritual homeland of Che Guevara, which makes it irresistible to so many visitors, even US citizens. Further, both Jamaica and Puerto Rico are English-speaking and the endemics can nearly all be very easily self-found during a relatively short stay. It is far harder to see all of the Dominican Republic's endemics, especially in one trip, and furthermore the country's infrastructure is such that the main birding area is remote and comparatively difficult of access. Very early morning starts will be essential to give you any chance of finding the majority of the most prized endemics. The upsides, however, make a visit well worth the challenge. The island's scenery is, in our opinion, some of the most dramatic in the region covered by this book, and the birding some of the very best. In part because of the relative lack of birding visitors, many of the important discoveries about the country's avifauna have been made by a handful of dedicated North American visitors, and local residents, during the last two decades. A valuable source of birding information about the country, organised by current resident, Steve Brauning, is at www.geocites.com/birdsofhispaniola/. These web pages are replete with links to local guides and the sounds of some endemics available to download, along with much other information.

Most visitors enter the Dominican Republic at Santo Domingo, and many endemics can be seen there or reasonably nearby. However, to find the majority of the more localised species you will need to drive to the coastal town of Barahona in the extreme south-west of the country, close to the Haitian border. This region is important because this is where the largest concentration of Hispaniolan endemics is found. Specialties of this part of the country include: Bay-breasted Cuckoo, Golden Swallow (now almost certainly endemic to Hispaniola), Hispaniolan Crossbill, Least Poorwill, Hispaniolan Nightjar, La Selle Thrush, Antillean Siskin, Green-tailed Ground Tanager (until recently considered to be a warbler), Hispaniolan Highland Tanager (also treated within the Parulidae until very recently), Antillean Piculet, Flat-billed Vireo, and both Eastern and Western Chat-Tanagers. However, Ridgway's Hawk is now probably already extinct in this region of the country, making it difficult to guarantee all of the endemics in a single trip. It is only 200km from Santo Domingo to Barahona, but the journey, largely along the Carretera Sánchez (Highway 2), takes 3–4 hours due to the abundance of small villages, and the presence of almost innumerable 'sleeping policemen' (huge speed bumps) and storm drains, which form 'trenches' in the road.

If you are on a dedicated birding trip, and can find cheaper flights to either Puerto Plata or Punta Cana airports, it is still possible to reach the main birding area in the south-west, but you will need to allot more time for travelling. If you are on a family holiday and intend to stay in either Santo Domingo or in one of the many tourist resorts along the eastern or northern coasts, some of the following sites may prove more accessible.

*Map of Dominican Republic*

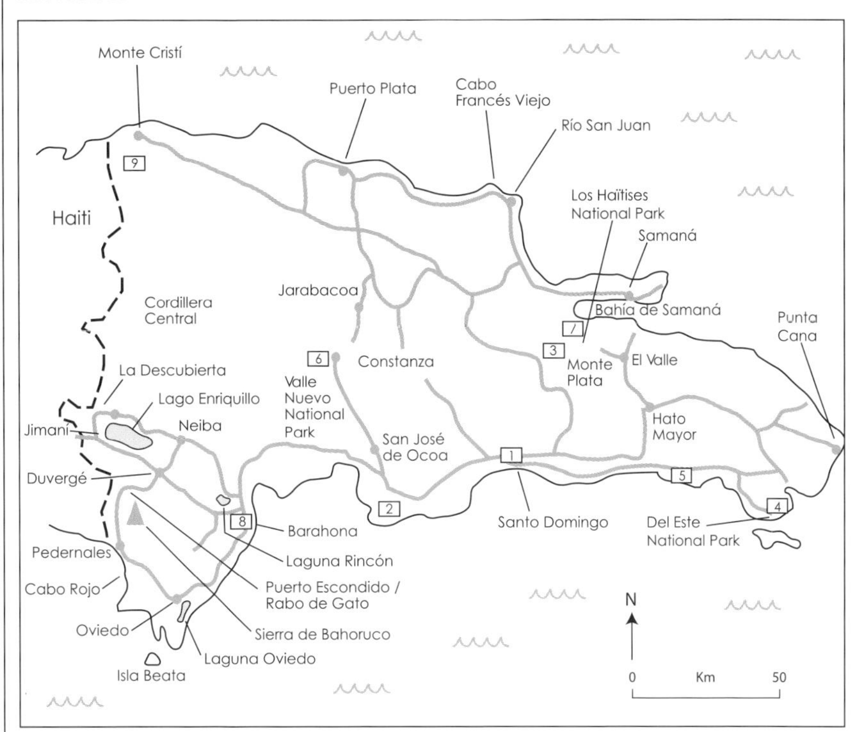


# Around Santo Domingo and the East

## National Botanical Garden (Jardín Botánico Nacional) *(See Site 1 on Dominican Republic map)*

If you are 'stuck' in the country's vibrant capital with time to kill, or on a (largely) non-birding vacation, then this 240ha site close to downtown is a remarkably peaceful place to visit. You can easily pass most of a morning here familiarising yourself with some of the commoner species. Although you will find the island's endemic bird family, represented by a single species, the Palmchat, to be ubiquitous throughout most the city's treed plazas, a visit to the botanic garden should reveal several other endemics.

*Location*

You may find it easier to take a taxi, but those with their own car will not find the area difficult to get to. To reach the gardens from Santo Domingo's seafront (the Malecón), proceed north on Avenida Abraham Lincoln, cross under the Avenida John F. Kennedy overpass, and continue to a roundabout. Keep in the right-hand lane and then take the first right exit, on Avenida de los Proceres. The garden is less than 1km further, with the actual entrance being reached by turning right at the next roundabout onto Avenida República de Colombia. There are a number of paved trails through the garden, but from the entrance it is best to go directly to the wooded area at the base of the plaza, while to reach the stream follow signs for Gran Cañada.

*Accommodation*

As in virtually any capital city, a wide range of hotels is available, to suit the vast majority of pockets. There are many pleasant options in the so-called Zona Colonial (the old city), where most 'regular' tourists prefer to stay. For the budget-conscious, in this part of the city, we have heard good reports concerning Bettye's Guest House, which is located near the junction of General Luperón and Arzobispo Meriño (telephone: 809-688-7649; e-mail: bettyemarshall@hotmail.com), and, close to the old city, the Hostal Plaza Yu, a few blocks inland of the Malecón, at Pasteur 208, esq. Josefa Perdomo (telephones: 809-686-7322; 685-2092; 688-7791; e-mail: hostalplazayu@yahoo.com; website: www.plazayu.com) can also be recommended.

*Strategy*

Birders may be allowed into the gardens as early as 06.30 (make sure you wear your binoculars), but the normal opening hours are 09.00 to 18.00. You may well need to visit the previous day (or telephone 809-567-6211) to organise early access. There is a small entry fee. This is a great site for West Indian Whistling Duck, which is best looked for along the stream and may be remarkably confiding. This is also a surprisingly productive site for the endemic Hispaniolan Parakeet, despite the species being generally quite difficult to find in the 'wilder' areas of the country. Wooded areas, especially those around the entry plaza, should be checked for migrant warblers in season.

*Birds*

The following endemics, near-endemics and other specialties can be seen here: West Indian Whistling Duck, Least Grebe, Limpkin, Zenaida and White-winged Doves, Hispaniolan Parakeet, Hispaniolan Lizard and Mangrove Cuckoos, Antillean Palm Swift, Vervain Hummingbird, Antillean Mango, Broad-billed Tody, Hispaniolan Woodpecker, Grey Kingbird, Red-legged Thrush, Black-whiskered Vireo (mainly summer only), Palmchat, and Black-crowned Palm Tanager.

## Salinas de Baní *(See Site 2 on Dominican Republic map)*

This area makes a reasonable site to day-trip from the country's capital, if based there, or a convenient stopover to and from the extreme south-west of the Dominican Republic. The saline lagoons are prime shorebird hunting grounds, especially during migration periods and in winter, and as a result have hosted quite a number of island (and Caribbean) rarities over the years, but for the endemic-hungry birder are perhaps of greatest interest for the occasional sightings of Black-capped Petrel that have been made from the Punta Salina during the species' winter breeding season.

*Location*

In Santo Domingo take Avenida 27 de Febrero west to Highway 2, which skirts the town of San Cristóbal, and shortly afterwards becomes single lane in each direction. Turn left in the town of Baní on the minor road signed to Matanzas and Salinas. If coming from the west (and Barahona), the turn-off is on the right before crossing the Rio Baní. After 22km, and approximately 90 minutes driving from the outskirts of Santo Domingo, you will see extensive mudflats and sand dunes, just as you reach the village of Las Calderas, where there is also a naval base. Continue on the road to Salinas village, just beyond which the road forks. The dirt road to the left shortly terminates in a parking area, from where you can explore an extensive area of saltpans and a narrow dyke can be followed as far as the beach. The right fork continues as a paved road to another area of saltpans.

***Accommodation*** Most visitors are likely to only visit the area en route to and from Santo Domingo and Barahona, or as a day trip from the capital, but there is at least one, rather unprepossessing hotel in Baní itself. There is a restaurant in Salinas.

***Strategy*** The area is very open and is probably best visited either early or late in the day, when the temperature should be cooler and there should also be less heat haze. A telescope will be extremely useful, and essential if you plan to look for Black-capped Petrels, or terns, from the beach. If you do visit at other times, make sure you have a hat, sunscreen and plenty of water. Should you have time on the way back, between Salinas and Baní turn left on the dirt road just before Matanzas, and follow this to the main highway (2). From here go straight across, on the dirt road heading north into the hills. After about 30 minutes the road peters out in fairly dry but disturbed forest, where you can find Yellow-billed Cuckoo, Hispaniolan Nightjar (at dusk), Antillean Piculet, Antillean Euphonia and Greater Antillean Bullfinch.

***Birds*** Arguably the star attraction is the chance of Black-capped Petrel, which has occasionally been seen from shore between November and April. Sightings are surely not guaranteed, but as yet very few people appeared to have tried their luck, so this globally threatened species, which nests in the high mountains on the Haiti border, may be more regular offshore in Ocoa Bay than is presently known. An evening watch, when petrels in general come closer inshore, is likely to be far the most productive. Nesting species include Snowy and Wilson's Plovers (both regular), Least Tern, and Willet, while other waterbirds typical of the site include the following: Reddish Egret, Tricoloured Heron, Clapper Rail, Killdeer, Least, Semipalmated and Western Sandpipers, both yellowlegs, Short-billed Dowitcher, and Royal Tern. There are few landbirds, but Burrowing Owl, Broad-billed Tody, Yellow Warbler (in areas of mangrove) and Black-crowned Palm Tanager are present. Rarities on the shore and pans have included several species of gulls and Red-necked Phalarope.

**OTHER SITES** If you are in the vicinity, in the capital Santo Domingo it is worth visiting the **Hotel Embajador** in the hour or so prior to dusk to observe the roost of Hispaniolan Parakeets, which sometimes numbers in excess of 100 birds, in the adjacent palm trees. During the day, the parakeets feed in the adjacent parkland. Occasionally, an individual of the recently introduced Olive-throated Parakeet is also seen at the roost, although there do not appear to be any recent records of this species from elsewhere within the city of Santo Domingo itself. Another sunset possibility, if visiting between May and August, is Antillean Nighthawk.

The beautiful White-tailed Tropicbird is suspected to breed (May to August) on the cliffs close to the main city centre of Santo Domingo, east of the Río Ozama; follow the coastal route, the Avenida España, and check the cliffs east of the **Acuario Nacional**.

**Monte Plata** *(See Site 3 on Dominican Republic map)* is a good area for both the thick-knee and Ashy-faced Owl. The countryside is characterised by agricultural land interspersed by palms. As ever, the owls are in the palms, but be aware that both species, Barn and Ashy-faced, are here. Thick-knees should be looked for in the agricultural fields and along dirt roads at night. Monte Plata is virtually due north of Santo Domingo, and is reached via Highway 11 and then by turning right when you reach the intersection with Highway 13.

## Del Este National Park *(See Site 4 on Dominican Republic map)*

This 310km² national park would make an excellent focus if your holiday is restricted to one of the huge resorts on the extreme eastern coast (indeed, there are even a couple of resorts right on the 'doorstep', just outside Bayahibe). If you are staying in one of the many resorts nearby, this site may be good for quick getaways and returns, permitting a sampling of the Hispaniolan endemics. The park, which was recently also designated an Important Bird Area, is the most visited protected area in the country, with an estimated 260,000 tourists annually.

***Location***

To reach the national park from Santo Domingo, head east on the coast road (Highway 3) through San Pedro de Macorís, then La Romana. Despite this being one of the country's main roads, traffic is not fast moving, and the journey will usually take at least three hours. Just after crossing the Río Chavón turn right on the road to Boca de Chavón and then follow the main road onwards to the coastal village of Bayahibe, where the park office is located. Approaching from Punta Cana and its holiday resorts, take the main road to Higüey, then head south towards San Rafael del Yuma. After 12km turn left towards Santo Domingo and San Pedro de Macorís. Turn off Highway 3 left, at the same point, just before the bridge over the Río Chavón. At the national park's office at Bayahibe, you need to pay the small entrance fee, which permits access to the area and the walking trail (and purchase tickets for the boat tour to Isla Saona, if you wish, although it is not an essential destination for birders). Guides can be hired through EcoParque, if needs be, but are not essential. To reach the park itself take the road towards the Club Dominicus Beach resort complex and after the last hotel on the right, the Hotel Coral Canoa, follow the dirt road towards the beach. Then, turn left when you are parallel to the sea and continue straight ahead until you reach Guaraguao.

***Accommodation*** Assuming that you are not staying at the Club Dominicus Beach or Casa del Mar resorts, which lie just 5km and about 1km east of Bayahibe, then there are a few other, much more reasonably priced options in the town itself. Recommended is the beach-side Hotel Bayahibe (telephone: 809-707-3684; fax: 809-556-4513). Day-tripping the area from Santo Domingo is possible, but hardly to be recommended, as you would spend at least six hours driving, rather than birding. However, there are quite a number of other options for accommodation in both La Romana, 30km to the west and San Pedro de Marcorís, a further 37km towards Santo Domingo.

***Strategy*** The main habitat within the 42,825ha national park is dry forest, characterised by many island endemic trees, where the following bird species should particularly be sought: Hispaniolan Parrot, Antillean Piculet, Flat-billed Vireo and White-necked Crow (listen out for their strange calls), as well as a variety of other dry forest species. Just after the Guaraguao park entrance, there is a good trail on the left-hand side that can be followed on foot for 3km. White-necked Crow is a specialty of this area, and the first Pearly-eyed Thrasher to be found on mainland Hispaniola was mist-netted here in February 1984. There does not appear to have been any subsequent records, but visitors clearly should keep a lookout for this species. The trail is initially sandy but then traverses limestone as it rises slightly. At the end there is a cave, known as El Puente, where there is sometimes a roosting Ashy-faced Owl. If you have a four-wheel-drive vehicle, it may be possible to continue south and traverse the entire peninsula, returning via Granchorra and Martel to San Rafael del Yuma, which would make a worthwhile circuit if staying in the Punta Cana area. If you do this, check the Boca de Yuma area for Hispaniolan Parrot. Among 'regular' tourists Del Este National Park is best known as the starting point for boat trips to Saona Island, which is mainly a beach area (and it is possible to stay overnight with a local family; ask the boatmen). There is a large Magnificent Frigatebird colony (200 pairs) en route (best visited in February or March), on Las Calderas, which it is possible to visit on request with the national park, and Common Bottlenose Dolphins *Tursiops truncatus* are sometimes seen during the journey. Hawksbill Turtle *Eretmochelys imbricata* nests on some of the beaches in the national park.

***Birds*** The following species found in the national park and the nearby Cumayasa Canyon area (see below) will be of particular interest: Magnificent Frigatebird, Brown Booby, American Flamingo (at least formerly, there was a colony on Saona Island), Red-tailed Hawk, Double-striped Thick-knee, Brown Noddy, White-crowned Pigeon (mangroves), Plain Pigeon (inland areas), Hispaniolan Parrot (a programme has been established to release captive-bred birds in the park, prompted by the species' island-wide decline), Mangrove and Hispaniolan Lizard Cuckoos, Barn, Ashy-faced and Short-eared Owls,

Least Poorwill, Broad-billed Tody, Antillean Piculet, Hispaniolan Woodpecker, Loggerhead Kingbird, Stolid Flycatcher, Cave Swallow, Flat-billed and Black-whiskered Vireos, Red-legged and Bicknell's Thrushes (winter), White-necked Crow (also common on Saona Island), Greater Antillean Grackle and Greater Antillean Bullfinch.

Bananaquit

***Other wildlife***

Interesting mammals known from Del Este include such poorly known species as Hispaniolan Solenodon *Solenedon paradoxus* (globally Endangered) and Hispaniolan Hutia *Plagiodontia aedium* (Vulnerable; the only member of this West Indian endemic family that occurs on the island), although you will require incredible good fortune to see either. The nominate subspecies of the uncommon West Indian Manatee *Trichechus manatus* is also known from inshore waters of the park, but if you take the trip to Isla Saona then you have more chance of catching a glimpse of a school of Common Bottlenose Dolphins.

**OTHER SITES**

The nearby **Cumayasa Canyon** *(See Site 5 on Dominican Republic map)* is well worth exploring. From Highway 3, between La Romana and San Pedro de Macorís, there is a crossroads just west of the bridge, at the top of the hill. Turning north onto a gravel road, drive down to the river edge, from where it is possible to hike north along the bed over rough terrain, with no real trail; forest birds such as Antillean Piculet, Greater Antillean Bullfinch, Loggerhead Kingbird, Broad-billed Tody and Black-whiskered Vireo are possible. Flat-billed Vireo was recorded in this region until the 1960s, but apparently not recently. In the evening any suitable fields should be checked for Double-striped Thick-knee and various owls, including Ashy-faced Owl. Especially productive are sugarcane fields that have recently been cut, and the best period is usually January to July.

A nearby area also worth exploring is **Ramón Santana/Cochoprimo**, which has dry scrub forest and some small wetlands. Least Bittern and Limpkin are possible, and there are very outside chances of Black and Spotted Rails. The area is best reached by turning north on the new bypass just east of San Pedro de Macorís, then turning right on the old highway near the old sugar mill and railway crossing. Proceed through the town of Ramón Santana onto the dirt road, until you see a right turn with a large girder bridge over the Río Soco. The next 2km are worthy of scrutiny. Ashy-faced Owl is another possibility in this area.

Pearly-eyed Thrasher has recently colonised the extreme east of the island. Check dry scrub woodland anywhere in the vicinity of **Punta Cana** for this species. To date the thrasher has been found only within a 3km radius of the town, but the species does appear to be a very recent arrival here (since about 1999) and is perhaps still spreading. Hispaniolan Parrot and Double-striped Thick-knee have also been recorded in this area, along with a variety of migrants. If you are staying in this region, take the minor road south from Bávaro, or north from Punta Cana, to **Charca de Bávaro**, a lagoon where Masked Duck, White-cheeked Pintail, Northern Jacana and Caribbean Coot have all been reported.

## Ebano Verde Scientific Reserve *(See Site 6 on Dominican Republic map)*

This 37km$^2$ protected area is the most accessible site for two key endemics, Eastern Chat-Tanager and the recently split White-fronted Quail-Dove (from Grey-headed), although neither species can be described as common, and both are best looked for very early in the morning. The reserve, which spans the altitudinal range from 800m to almost 1,600m, is sited almost equidistant from the north and south coasts, and therefore equally accessible to those based either in Santo Domingo or Puerto Plata. However, the drive from Santo Domingo should take approximately 1.5 hours, but will probably take almost double this if coming from the north coast.

***Location***

To reach Ebano Verde Scientific Reserve from the south, drive north-west from Santo Domingo on the Autopista Duarte (Highway 1), and beyond Bonao take the left turn-off to Constanza, from where you will start climbing up a steep hill on a winding road (the Carretera de Casabito; Highway 12). If approaching from the north coast, search for signs for Highway 1 and Santo Domingo in the town of Santiago de los Caballeros, and follow this road to La Vega and beyond. The turn onto Highway 12 is located just beyond the village of Pringamosa. At the summit of the road, you will see the first entrance (Casabito) to the reserve, which entry point is only passable in a four-wheel-drive vehicle. This entry is signed close to a shrine at the summit near the radio masts.

***Accommodation***

If you are not day-tripping the site, the best place to stay in this area is the Hotel Jacaranda (telephone 809-525-5590), on the left side of the highway a few kilometres past the Bonao exit. It is also a truck stop, with a cafeteria and restaurant on-site. Alternatively, there are a couple of relatively basic (and cheap) hotels in Constanza, but reaching this town will involve driving about 25km further along the narrow and winding Highway 12. Another option is the Hotel Altocerro, just east of Constanza (telephone: 809-539-1553; e-mail: c.matias@codetel.net.do).

***Strategy***

Ebano Verde is a private reserve, originally created in 1989, and is gated, making it essential that you arrange your permit to visit at least two weeks in advance (telephone: 809-565-1422; e-mail: fund.progressio@codetel.net.do), especially if you need a Fundación Progresso guard to be available to open the upper gate. From here you initially climb uphill to the radio masts, and then down to the visitor centre. The main entrance (no gate) is further along Highway 12, after the second bridge on the right, El Arroyazo, while the Fernando Aquino Visitor Center is sited at the other end of the trail. Depending on which species you prefer to target will probably determine at which end of the trail you start. The 6km-long Arroyazo Sendero de Nubes trail passes initially through pine forest and second growth, then riparian forest dominated by a species of palm along the stream (look for White-fronted Quail-Dove in this area). Subsequently, the trail enters undisturbed cloud forest, where you should search for Eastern Chat-Tanager (as always, best in the first hour of daylight) and the quail-dove, before rising steeply and terminating at the Casabito entrance. Your best chance of finding a quail-dove is, as ever, to ensure that you move as quietly as possible along the trail, hoping to surprise one in the act of feeding or crossing an open area.

***Birds***

Endemics, near-endemics and other specialties of this fine reserve include: Red-tailed Hawk, Sharp-shinned Hawk (upper section), Scaly-naped Pigeon, Plain Pigeon (quite common along the road), White-winged Dove, White-fronted Quail-Dove, Hispaniolan Parrot (upper section), Hispaniolan Lizard Cuckoo, White-collared and Antillean Palm Swifts, Antillean Mango, Hispaniolan Emerald, Vervain Hummingbird, Broad-billed and Narrow-billed Todies, Hispaniolan Trogon (common), Antillean Piculet, Hispaniolan Woodpecker, Hispaniolan Pewee, Greater Antillean Elaenia, Stolid Flycatcher, Rufous-throated Solitaire, Red-legged Thrush, Hispaniolan Palm Crow (rare), Caribbean Martin, Golden Swallow (rare), Pine Warbler, Cape May Warbler (common in winter), Antillean Euphonia, Hispaniolan Spindalis, Black-crowned

Palm, Green-tailed Ground and Hispaniolan Highland Tanagers, Eastern Chat-Tanager (probably only likely to be seen using playback, especially in areas with dense undergrowth dominated by ferns), Antillean Siskin, Rufous-collared Sparrow (common in the upper section, at one of its few localities in the Caribbean), and the relatively ubiquitous Yellow-faced and Black-faced Grassquits.

**OTHER SITES**

Those with an exploratory bent may be well rewarded if they decide to visit the **Valle Nuevo National Park** (formerly Scientific Reserve), which covers the area south of the town of Constanza, on the road towards San José de Ocoa. The area is an extremely important water catchment area. From Constanza, take Highway 41 south through Colonia Japonesa to Valle Nuevo; the road is quite rough and a four-wheel-drive vehicle is recommended. Especially worth exploring is the track signed from the old army camp to Alto Bandera. Commoner birds along here include Antillean Palm Swift, Antillean Mango, Hispaniolan Emerald, Hispaniolan Woodpecker, Greater Antillean Elaenia, Red-legged Thrush, Pine Warbler, Hispaniolan Spindalis, Black-crowned Palm Tanager, Yellow-faced Grassquit, Antillean Siskin, Hispaniolan Crossbill and Rufous-collared Sparrow. Among the many interesting species in this region are White-fronted Quail-Dove, Hispaniolan Parakeet, Stygian Owl, Golden Swallow, Hispaniolan Palm Crow, La Selle Thrush, Hispaniolan Highland Tanager, and Eastern Chat-Tanager. Bicknell's Thrush occurs in winter, and interesting mammals include Hispaniolan Solenodon and Hispaniolan Hutia, although neither is likely to be seen by casual or even more determined visitors.

## Los Haïtises National Park *(See Site 7 on Dominican Republic map)*

Currently, this is the only regular site for Ridgway's Hawk. The 200km$^2$ protected area is one of the wettest parts of the island and comprises dense lowland broadleaf forest growing on karstic limestone, as well as mangrove. The protected area was initially declared as long ago as 1968, and national park status was bestowed in 1976; its boundaries were extended in 2000 and again in 2004. Forested hills of limestone, called mogotes, dot the topography. Once cleared for agriculture, the forest is regenerating due to protection, but accessing the park is difficult, which is three hours from Santo Domingo on the south side of Samaná Bay.

***Location***

From Santo Domingo, drive east past the international airport in the direction of San Pedro de Macorís (58km). On reaching the latter, take the northbound turn, on Highway 4, to Hato Mayor (33km); from the latter town, continue to Sabana de la Mar, on the minor road via El Valle, where the park office is located at the north end of town, close to the pier.

***Accommodation***

The best place to stay is probably Caño Hondo, which is obscurely signed on the left just as you enter Sabana de la Mar, and is approximately 1km from the boat dock. Contact details: website: www.paraisocanohondo.com; e-mail: canohondo@verizon.net.do; and telephones: 809-248-5995 or 809-696-3710. Meals can also be taken here. There are also a few hotels and restaurants in El Valle, but none can be especially recommended.

***Strategy*** The main reason for visiting this national park is the chance of finding the globally threatened and declining Ridgway's Hawk. There are two real possibilities for accessing the park area, which is not straightforward and either way it is preferable to hire the services of a guide, which can be done through the national park office. The easiest method is to hire a boat from Caño Hondo and explore the mangrove area along the Río Yabón, downstream towards the bay, and pausing to explore the various trails through the lowland forest. Reputedly the best area is Los Naranjos. The owner of Caño Hondo can also organise a guide to follow a two-hour or so trail that starts in the pasture near the lodge, and sometimes the whereabouts of a hawk nest may be known along here. The alternative option, which often produces sightings of Ridgway's Hawk, is to take the difficult-to-follow, overland trail from Trepada Alta, reached from El Valle (on the road between Sabana de a Mar and Hato Mayor), where you should ask directions. The road can be very rough and muddy. Some of the rangers at Trepada Alta are familiar with the hawk and can assist you for a small fee. Again, the rangers may be aware of a nest of the hawk. The trail is reputed to be both treacherous and slippery, and is lined with nettles.

Another area for Ridgway's Hawk is Los Limones, which lies inland on the western edge of the park, and can be accessed via the town of Sabana Grande de Boyá, although it is apparently a rough and long ride to the birding area. At present, there is a Peregrine Fund nest-monitoring project in progress there. If you intend to visit, you should definitely contact Lance Woolaver (e-mail: lancewoolaver@hotmail.com) in advance, as he might be able to help you arrange access. The whole area is hot and very humid, and you should definitely carry plenty of water anywhere in the park.

***Birds*** Besides Ridgway's Hawk other species you may see include: Least Grebe, Brown Booby (which nests offshore), Magnificent Frigatebird (also nests offshore), Red-tailed Hawk, Limpkin, Hispaniolan Parrot, the globally threatened Plain Pigeon, White-crowned Pigeon, Ruddy Quail-Dove, Hispaniolan Parrot, Hispaniolan Lizard and Mangrove Cuckoos, Least Poorwill, Antillean Palm Swift, Antillean Mango, Hispaniolan Emerald, Vervain Hummingbird, Broad-billed Tody, Antillean Piculet, Hispaniolan Woodpecker, Hispaniolan Pewee (uncommon), Stolid Flycatcher, Grey Kingbird, White-necked Crow (the park is one of the strongholds of the species, and it is easily seen around Caño Hondo), Palmchat, Black-whiskered Vireo, Black-crowned Palm Tanager, the introduced Village Weaver, Greater Antillean Grackle, and the recently split Hispaniolan Oriole. In season, the area can be very productive for wintering North American warblers, and Bicknell's Thrush also occurs in winter.

# The South-west

*Map of SW Dominican Republic*

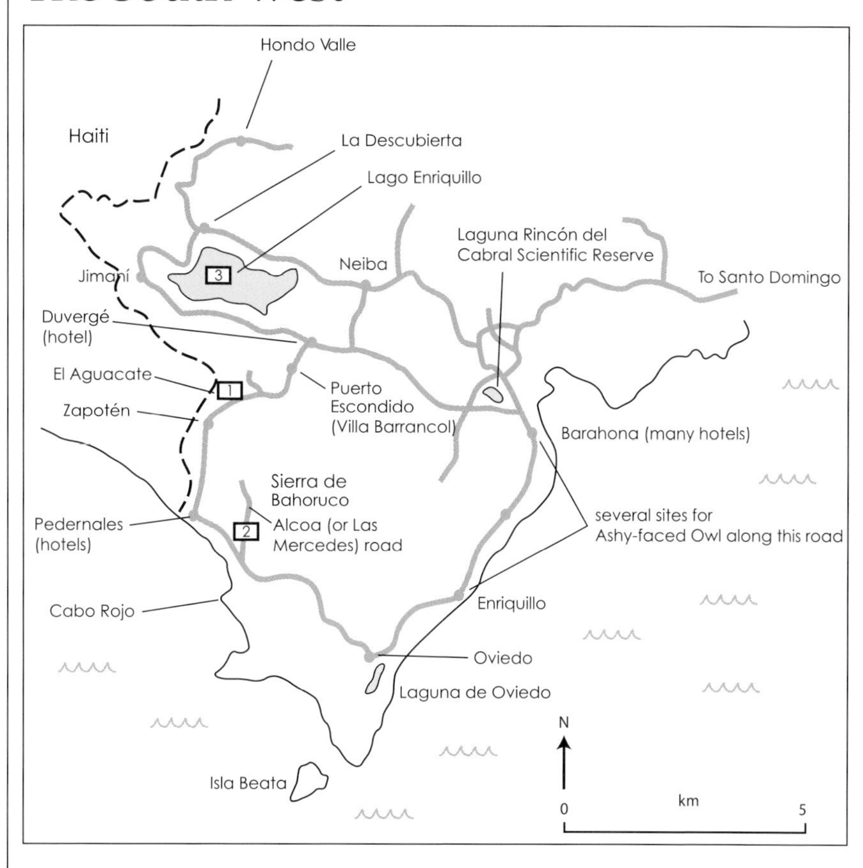


## Barahona *(See Site 8 on Dominican Republic map)*

The gateway to the most important sites in the Dominican Republic is Barahona, located on the west side of Neiba Bay, in the south-west of the country, which has served as a base for many birders determined to see the Hispaniolan endemics over the years.

*Location*

Barahona is a 200km drive from Santa Domingo, via Highway 2 and Highway 44, but it takes close to four hours to navigate due to the never-ending small villages en route and huge 'sleeping policemen' (speed bumps).

*Accommodation*

As mentioned in the introduction to the country, the major disadvantage to birding this area is that, for groups and those that like their 'creature comforts', until recently the only satisfactory lodging was in Barahona. If you are intending to arrive at a weekend or during a holiday period, then you would be well advised to attempt to book in advance. Consistently recommended and used by birders have been the mid-range Hotel Caribe (telephone: 809-524-4111), San Martín (telephone: 809-524-5821) and the Guarocuya Hotel (telephone: 809-524-2211), right on the beach at the south end of town, but there are both cheaper and more salubrious options, depending on your budget and taste. There is a reasonable range of restaurants in this seaside town. However, if you are travelling in a small group and want to stay an hour closer to at least some (but not all) of the birding sites, then there is at least one small hotel in the town of the Duvergé, at the south-east end of Lago Enriquillo. This is the cheap and relatively basic Hotel Ana (no off-road parking, but there is a well-recommended family-run restaurant across the street). Heading west on the main street through Duvergé, go right at the Shell station just beyond the Zona Militar. The restaurant (very small) is two blocks down here at the corner of General Cabral and Damiro Ademes. There is also a small hotel in Cabral. The best option at present is probably to stay at the basic but clean campsite, Villa Barrancol, established by Kate Wallace of Tody Tours just beyond Puerto Escondido. Villa Barrancol, which is at the head of the Sendero Ecológica Rabo de Gato, is well signed within the village; instead of turning right towards El Aguacate in the centre of the village, turn left, and then right at a shallow fork in the road, go past a dam on the right and then take the next right turn. Continue along here for about 1km until the campsite appears on the left. Villa Barrancol has two small chalets, each with two beds, as well as camping equipment. It is also possible to arrange for women from the local village to cook for you. If you intend to stay at the camp, it is probably best to arrange this in advance with either Kate Wallace (e-mail: todytours@gmail.com) or local birding guide Miguel Landestoy (mango_land@yahoo.com). Another option might be to camp in the mountains, but this should be done with caution and with the permission of any local people in the area. At the time of writing (early 2008), Pedernales only had about three hotels, of which two were very basic, but the seaside Hotel Villa del Mar (telephone: 809-524-0448; e-mail: ferreras_victor@yahoo.es), off to the left at the entrance to the town, has been recommended recently by some birders, and makes a fine base to visit the Alcoa Road and Cabo Rojo.

***Strategy*** From Barahona there are three major areas to cover, any one of which will require a long drive to reach, necessitating a very early start, as you will need an entire day to find the birds (some of which are only readily located at either dusk or dawn) and then return to the hotel at a decent hour.

## A – Sierra de Bahoruco–El Aguacate

*(See Site 1 on South-west Dominican Republic map)*

To get the most from any of these sites, especially those from La Placa to Zapotén, really requires that you be on site at dawn or shortly afterwards. Therefore a starting time of 03.00–04.00 from Barahona is highly recommended, although when covering the north slope of the Sierra de Bahoruco you can shave an hour off this, if you elect to stay in Duvergé or at Villa Barrancol.

From Barahona, drive north on Highway 44 to the first major junction in the road, where you should turn left to Duvurgé and Jimaní. From here it is just over 50km to Duvergé, where you should turn left (south) towards Puerto Escondido on Calle Carmen, which is the first street west of the telecommunications office and a few blocks west of the Shell station. Just before the road becomes dirt, there is a yellow sign for the Sierra de Bahoruco National Park, although the sign faces a side street and is difficult to see. Continue uphill (keeping a lookout for Olive-throated Parakeet if it is daylight), passing a hydroelectric plant on the right, towards the small town of Puerto Escondido (8km), which is reached after approximately half an hour and will be visible from the crest of the hill. There are often Burrowing Owls along this road bank and indeed along the road beyond Puerto Escondido. The dry scrub at the crest of the hill can also be usefully searched for Flat-billed Vireo, but remember that you are still some distance from your main destination. On arriving in Puerto Escondido you will come to a T-intersection, from where there are two options. Just before this, on the right, is the national park office, which is the building on the corner of the first junction in the village, where you should (depending on the time of day) stop and pay the small entrance fee.

To access the area known as **Rabo de Gato**, turn left at the junction and continue a short distance before turning right at a Rabo de Gato sign. Go through a shallow dip, cross the canal and finally turn right at a second Rabo de Gato sign; you will see a fence and gate on the left through which lies Villa Barrancol (see above). Beyond the campsite, there is a relatively easy wide trail through riparian habitat, which can produce both tody species, as well as Hispaniolan Lizard Cuckoo, Hispaniolan Trogon, Antillean Piculet, White-necked Crow (check palms fringing any of the irrigation channels in the area), Flat-billed Vireo, Antillean Siskin (which feeds in the large fig trees), Hispaniolan Oriole, and possibly even Bay-breasted Cuckoo. The latter is best searched for between the picnic tables on the left-hand side (where there is also a sign with a Palmchat and Hispaniolan Trogon on it) and another sign announcing the end of the trail. This area is also productive for quail-doves, with White-fronted, Key West and Ruddy all possible. After dark, Northern Potoo can be found in this area, as can Least Poorwill (especially around the campsite) and occasionally even Ashy-faced Owl. At the point where the trail 'ends', there is in fact a fork. To spend longer birding this area, you could continue to the right and then follow the trail as it bends to the left, and continue until you reach an ungated fence, immediately beyond which is another fork. By going left, following the bed of a dry stream, you reach an area with tall trees on both sides of a valley. This is another known spot for Bay-breasted Cuckoo. If taking the right-hand trail at the fork, you will pass a farmstead on the left before eventually reaching a very large cleared area, where the trail effectively ends. The area between the small holding and the clearing is another spot for the cuckoo.

Back in Puerto Escondido, by turning right at the T-junction you will be taking the road to El Aguacate, and **Zapotén**. As the agricultural area on the left-hand side, which can be productive for Antillean Nighthawk in summer (especially over the irrigated fields), becomes dry broadleaf forest, and just before the road begins to climb, search again for Bay-breasted Cuckoo, Flat-billed Vireo, Antillean Piculet, and Plain Pigeon (which can also be found in dry forest along the road between Cabral and Duvergé). There may also be wintering warblers in this area (especially Prairie, Cape May and Black-throated Blue), as well as plenty of Green-tailed Ground Tanagers, and at either last light or just before dawn there are good chances for Northern Potoo, Least Poorwill (which sometimes continue to vocalise for the first hour of daylight), and Hispaniolan Nightjar. Some 10km from Puerto Escondido, you will see a large yellow sign, 'La Placa'; between here and Los Naranjos, about 5km further on, is typically the best area for Bay-breasted Cuckoo, and there is a side trail off to the right opposite the sign, which can be particularly good. The next stretch through more humid forest sees a change in avifauna, and the road become decidedly more difficult, as it is very rutted and principally consists of loose pebbles and larger rocks (a four-wheel-drive is not essential, but unless you are a confident driver, you may find it very difficult going without such a vehicle, and you need high clearance). At this point you are actually ascending a dry riverbed. Hispaniolan Trogon is common along here, and there are still chances for Key West Quail-Dove and more opportunities for Bay-breasted Cuckoo.

Eventually, you will reach the military post at El Aguacate, more or less on the border with Haiti (note the stark contrast where the forest begins and ends). Inform the soldiers of your purpose and they will record your license plate and wave you on, although a few cigarettes (not packs) will be appreciated, and serves to promote goodwill. We have heard that some birders have been informed that they needed a soldier escort to accompany them in recent years. Grey-crowned Palm Tanager has been claimed from the gully just before you reach the checkpoint, but none of the Dominican Republic claims of this species, which is otherwise endemic to neighbouring Haiti, has been documented, and considerable doubt must exist concerning most such sightings.

The wet broadleaf forest transitions into pine forest after a further 5km as the road winds along the ridge. The first area where pines predominate, about 15 minutes drive past the checkpoint, is a good place for Hispaniolan Nightjar predawn. Beyond this is the area to search for Western Chat-Tanager and La Selle Thrush. More sightings of Grey-crowned Palm Tanager have been claimed from this area, but observers should bear in mind the greater likelihood of young Black-crowned. The thrush is easiest at dawn and dusk (when they sing), and was until recently difficult (probably exclusively a matter of good luck) during daylight hours. One of the best areas for both species is that marked by a national park sign with, appropriately enough, a picture of a La Selle Thrush. As of spring 2009, both Western Chat-Tanager and La Selle Thrush were relatively easy to find in this area during the first couple of hours of daylight, along with Hispaniolan Highland Tanager and many other endemics, as well as wintering Bicknell's Thrushes, also to be found here. Zapotén is the area where the Haitian potato market now stands, and the surrounding vegetation has to some extent been cleared. Indeed parts of the area bear only a sad resemblance to their still relatively pristine state in the early 1990s, but ironically this might have made some of the more difficult and skulking birds easier to locate. Only scraps if anything of the bulldozer, mentioned in countless trip reports, remain. There is at least one trail leading uphill away from the road, just beyond the first of three clearings, on the first left-hand hairpin bend above the potato market. Further on you will see another sign ('Observación de aves') announcing a turn to the right, which continues to a clearing in the pines, around which it is possible to find Hispaniolan Crossbill, Antillean Siskin, Pine Warbler and Hispaniolan Trogon. Above this clearing, on a clear day, you can see the cliffs of the Loma del Toro where Black-capped Petrel breeds.

While it is possible (sometimes with difficulty, occasionally not at all) to continue across the mountains to the town of Pedernales, on the south coast, there is plenty to search for here and you could easily spend all day along the road between the highest point in the road (at well over 2,000m) and Puerto Escondido. If you decide to continue to Pedernales, it is doubtful that you will encounter any additional species, although there have been recent claimed sightings of Ridgway's Hawk. Furthermore, there have been occasional records of La Selle Thrush and Western Chat-Tanager (which like the Eastern species prefers areas of dense undergrowth and, like the thrush, is far more easily heard than seen) between Los Arroyos and Mencia.

***Birds***

This area is one of the most productive for birds on the entire island and the following are all possible, although some species are distinctly difficult. Black-capped Petrel (winter-breeder; it would probably be necessary to camp in the mountains to hear birds visiting the colony, and in any case most breed in adjacent parts of Haiti), Red-tailed and Sharp-shinned Hawks, Scaly-naped and White-crowned Pigeons, Zenaida and White-winged Doves, Key West Quail-Dove (uncommon; best looked for in drier areas), White-fronted Quail-Dove (perhaps declining or at least more difficult to see in recent years due to increasing disturbance), Ruddy Quail-Dove, Olive-throated Parakeet (introduced but seemingly well established), Hispaniolan Parakeet, Hispaniolan Parrot (fairly common still), Hispaniolan Lizard and Bay-breasted Cuckoos (the latter is one of the most difficult endemics), Hispaniolan Emerald, Vervain Hummingbird, White-collared and Antillean Palm Swifts, Ashy-faced Owl (rare), Burrowing Owl, Antillean Nighthawk (summer), Chuck-will's-widow (winter), Least Poorwill (uncommon), Hispaniolan Nightjar (common), Hispaniolan Trogon, Broad-billed and Narrow-billed Todies, Hispaniolan Woodpecker, Hispaniolan Pewee, Greater Antillean Elaenia, Stolid Flycatcher, Loggerhead Kingbird (uncommon), Grey Kingbird, Caribbean Martin, Golden Swallow (rare but regularly seen), White-necked Crow, Hispaniolan Palm Crow (both crows are rather uncommon), Red-legged and La Selle Thrushes, Rufous-throated Solitaire (far more easily heard than seen), Flat-billed and Black-whiskered Vireos, Palmchat, Pine Warbler, Bananaquit, Antillean Euphonia, Antillean Siskin, Hispaniolan Crossbill (a recent split from White-winged Crossbill), Black-crowned Palm and Green-tailed Ground Tanagers, Western Chat-Tanager, Hispaniolan Highland Tanager (usually rather uncommon), Hispaniolan Spindalis, Greater Antillean Bullfinch, Greater Antillean Grackle and Hispaniolan Oriole.

## B – Alcoa Road *(See Site 2 on South-west Dominican Republic map)*

Highway 44 begins at Barahona and terminates in Pedernales. This is a 125km, two-hour drive on an excellent highway, which follows the southern coastline and goes south-west and inland north of Lago Oviedo. At a point 48km from the town of Oviedo, and just beyond signs to Bahía de las Aguilas, there is a bridge. Immediately after this take the right turn, and then go left (by continuing right, you will head for the coast at Cabo Rojo). You are now on the Alcoa (or Las Mercedes) road. The paved road climbs through dry broadleaf forest (which could be searched for both Hispaniolan Nightjar and Least Poorwill at dusk) until it reaches the Hoyo de Pelempito ticket station, at the town of Las Mercedes, beyond which is an area known as Aceitillar.

Above Las Mercedes there is a very wide hairpin turn. This is a good site for Western Chat-Tanager (to date, the lowest known for this species), in the shrubby growth on the right side of the turn. Many other endemics are also possible in the immediate area. Just beyond here is the abandoned Alcoa bauxite mine. As you enter the area, which is flanked by pine trees, you will see a concrete pond on the right (called La Charca), where both Hispaniolan Crossbill and Antillean Siskin come to drink. If you spend some time here you should have a very good chance of seeing both species, making this arguably the most reliable site on the island for these birds. Just slightly further on, a former mine working is often a good area to find Golden Swallows hawking insects. Provided you have purchased a ticket and have a reasonably high-clearance vehicle, it is possible to continue to Pueblo Viejo and the new visitor's center at Hoyo de Pelembito (the end of the road), which has an overlook that is good for viewing parrots flying to roost in the evening. Hispaniolan Trogon is very common in this area, and the trails at Pueblo Viejo can be productive for both Western Chat-Tanager and Bicknell's Thrush, which winters in large numbers in the area. However, seeing one without using tape playback of the winter call is likely to be very difficult.

***Birds***

Another very productive and generally very accessible area (although still a long drive from Barahona), the following are all possible, although again some species are distinctly difficult: Red-tailed and Sharp-shinned Hawks, Northern Bobwhite, Scaly-naped and White-crowned Pigeons, Plain Pigeon (regularly seen above Las Mercedes and at Pueblo Viejo), Zenaida and White-winged Doves, White-fronted Quail-Dove (Pueblo Viejo), Olive-throated Parakeet (introduced but seemingly well established), Hispaniolan Parakeet (a good area), Hispaniolan Parrot (fairly common still), Mangrove and Hispaniolan Lizard Cuckoos, Antillean Mango, Hispaniolan Emerald, Vervain Hummingbird, Antillean Palm Swift, Ashy-faced Owl (most frequently heard or seen near Hoyo de Pelembito), Burrowing Owl, Stygian Owl (has been heard in the vicinity of the drinking pool), Antillean Nighthawk (summer), Least Poorwill (uncommon), Hispaniolan Nightjar (common), Hispaniolan Trogon, Broad-billed and Narrow-billed Todies, Hispaniolan Woodpecker, Hispaniolan Pewee, Greater Antillean Elaenia, Stolid Flycatcher, Grey and Loggerhead Kingbirds, Caribbean Martin, Golden Swallow (regular around the old bauxite workings), White-necked Crow, Hispaniolan Palm Crow (both crows are rather uncommon but regularly seen), Bicknell's and Red-legged Thrushes, La Selle Thrush (common in the higher parts well above Las Mercedes), Rufous-throated Solitaire (far more easily heard than seen), Palmchat, Pine Warbler, Bananaquit, Antillean Euphonia, Antillean Siskin, Hispaniolan Crossbill (a recent split from White-winged Crossbill), Black-crowned Palm and Green-tailed Ground Tanagers, Western Chat-Tanager, Hispaniolan Highland Tanager (usually rather uncommon), Hispaniolan Spindalis, Greater Antillean Bullfinch, Greater Antillean Grackle and Hispaniolan Oriole.

Hispaniolan Lizard Cuckoo

If you have time on the return journey, you can cross the main highway, and proceed to the coast and **Cabo Rojo**, where a small wetland on the left-hand side of the road attracts many herons, shorebirds (including more interesting species such as Short-billed Dowitcher and Stilt Sandpiper) and ducks such as Blue-winged Teal and White-cheeked Pintail. West Indian Whistling Duck and Antillean Nighthawk (summer) have occasionally been seen here at dusk. Mangrove scrub harbours an endemic subspecies of Yellow Warbler and from time to time many wintering warblers, as well as other residents such as Antillean Mango and Broad-billed Tody, while along the shore you may see terns, herons and Brown Pelican. The whole area can be very good for migrants, but to connect with a good day obviously depends on luck. There is also breeding White-tailed Tropicbird, Caribbean Martin, Cave Swallow and Antillean Palm Swift (which here breed in caves) in this area, but gaining access to the cliffs at the end of the point, via the dock area that serves the mine, is usually difficult.

Further along, a quick stop at the very shallow and brackish **Oviedo Lagoon** (a proposed Ramsar site and part of the 165,000ha Jaragua National Park) may prove productive for American Flamingo, Roseate Spoonbill, White Ibis and other waterbirds. An observation tower has recently been constructed here, but most of the rest of the area is usually difficult to access, unless you can organise a boat trip. White-cheeked Pintail and West Indian Whistling Duck are again possible here, as is White-necked Crow in the large palms. Roadside pools between Oviedo and Barahona can also be productive for shorebirds, especially during passage periods, although Cabo Rojo is likely to be better.

## C – Lago Enriquillo and environs *(See Site 3 on South-west Dominican Republic map)*

This highly saline lake has the distinction of lying below sea level and, at over 40,000ha, is the largest lake in the Caribbean region. It is a National Park and Ramsar site, and is easily visited from Barahona, perhaps as an afternoon trip after a full day at either of sites A and B, or if you stay in Duvergé then you are on the 'doorstep'. As for site A, leave Barahona west on Highway 46 through Cabral and continue onward to Duvergé. After passing Duvergé you will soon see Lago Enriquillo on the right. Alternatively, about 5km before Duvergé there is a right turning (marked by a large Indian statue) to Neiba, where you can turn left and continue around the north side of the lake. By taking either route, north or south, and by using a telescope it should be possible to find American Flamingo, along with a variety of herons, terns (including Caspian Tern) and shorebirds. Most stands of palms around the lake are, perhaps surprisingly, not good sites for Hispaniolan Palm Crow (although it does occur on Isla Cabritos, an island and the original site of the national park, in the middle of the lake, which can be accessed by boat, arranged from the national park office in La Descubierta). However, the palm crow can also be found along the south shore of the lake, about 5km beyond (west) of the town of Las Baitoas, where there is a dirt road that leads off to the right on a bend in Highway 46. Drive down this road checking (and listening) for crows. On the north side of the lake, at La Descubierta, it is possible to regularly find White-necked Crow just east and south of the town. Check the trees around the restaurant at the end of the town on the south side of the road, or take any one of the narrow dirt roads, e.g. signed Las Barias, that lead towards the lake, checking the stands of palm trees. White-necked Crow can also be found near the public swimming area, with its turquoise-green building, on the north side of

Highway 46 on the south shore of the lake. Turn off here and continue on the dirt road past the swimming area until you reach a fork on the road. Walk right and check the palms for crows. Just a few kilometres to the west of here Hispaniolan Palm Crow can be found in the tall palms right beside highway 46, along with Plain Pigeon. Dry scrubby woodland between La Descubierta and Jimaní might be productively searched for Least Poorwill, although the species is far from common in this region of the country. The whole area of the Lago Enriquillo is a reasonable one to find the globally threatened Plain Pigeon, and the rare American Crocodile *Crocodylus acutus* is sometimes seen along the shore of the lake.

It is possible to ascend the lower slopes of the Sierra de Neiba from the north side of the lake, for instance by turning right along the minor road (Highway 47) towards Hondo Valle just west of La Descubierta. At least some of the same species of the Sierra de Bahoruco can be found here, and Stygian Owl has been reported from the Cueva de la Mulcielagos, near Los Pinos, the first village along this road. However, the road is rather rough, narrow and has many twists and turns, and good habitat is difficult to access.

***Birds***

The area is productive for a relatively small number of species, with the particular highlights being the two species of crows. The following are all possible: American Flamingo, Plain Pigeon, White-winged Dove, Hispaniolan Parakeet, Hispaniolan Parrot, Mangrove and Hispaniolan Lizard Cuckoos, Burrowing Owl, Antillean Nighthawk, Hispaniolan Nightjar, Antillean Mango, Antillean Palm Swift, Broad-billed Tody, Antillean Piculet, Hispaniolan Woodpecker, Stolid Flycatcher, Grey Kingbird, Palmchat, Hispaniolan Palm and White-necked Crows, Black-crowned Palm Tanager, Village Weaver, and Hispaniolan Oriole.

## OTHER SITES

There are a number of potential sites to find the generally uncommon Ashy-faced Owl in the vicinity of **Barahona**. Remember that Barn Owl (split from Old World birds by some authorities) occurs in this area too, so not every *Tyto* owl in this area is the endemic species. One good area to check is the dirt road that heads inland about 3km south of the town immediately before a checkpoint. Coming from the town, there is a sign for Santa Elena on the right-hand side of the road, while if coming from the south, the dirt road is immediately after the checkpoint and has a sign for the Catholic Technical University of Barahona beside it. From the main highway drive 2km and especially check the large palms. Least Poorwill and Northern Potoo are also regularly found in this area. Another possibility is the valley 4.3km south of the Hotel Caribe. Turn west on the tarmac road here, immediately beyond an advertisement board on the same side of the road (the left), and bird the area from 2km further along. Northern Potoo, Least Poorwill and Key West and Ruddy Quail-Doves have also been seen in this area. Another locality for the owl is reached by turning west off the same road, this time at a locality known as Los Cacaos, in a ravine about 7.2km south of the Hotel Caribe. Park just beyond the house with the beehives and follow the trail into the ravine. The quail-doves and Northern Potoo have also been seen at this locality.

Also very close to Barahona is the **Laguna Rincón del Cabral Scientific Reserve**, an Important Bird Area and National Wildlife Refuge, which is best reached by taking the turn to Duvergé off Highway 44 north of Barahona, and continuing 10km to the small town of Cabral, where a minor road leads off north (right) to the village of El Peñón. Just after crossing the railway line you will see the reserve's headquarters, where you can hire a guide to explore the environs of the lake (which is the largest fresh waterbody in the country). Access to the lake is not necessarily easy and depends on locating and investigating any dirt roads leading off the minor road west and north in the direction of the lake (e.g. following drainage channels). Waterbirds such as Pied-billed Grebe, White-cheeked Pintail, Masked Duck, Caribbean Coot (exceptionally as many as 3,000 birds) and Northern Jacana are all possible, as is a variety of egrets and shorebirds, should you be successful in finding any muddy areas of shoreline. In winter other ducks can include American Wigeon (up to 10,000), Blue-winged Teal (up to 25,000), Lesser Scaup (up to 90,000) and Ruddy Duck (up to 10,000), but numbers are typically much lower than this, especially in recent decades, and remember that access difficulties mean that you are unlikely to see anything like such huge numbers of birds. The surrounding scrub has a typical range of passerines, including the introduced Nutmeg Mannikin, as well as the globally threatened Plain Pigeon.

Also heading towards Duvergé, again at the crossroads with the black cross in the centre of Cabral, turn left (south) and follow the sign to **Polo**. The road soon ascends through extensive scrub with the usual range of species in such habitat. Further on, this paved road passes through some moist riparian woodland and shade coffee plantations, with Scaly-naped Pigeon and Eastern Chat-Tanager (although the identity of these birds has been questioned and they were originally reported as being Western Chat-Tanager), as well as a range of other species typical of similar habitat elsewhere in the Sierra de Bahoruco (e.g. Hispaniolan Highland Tanager, Hispaniolan Parrot, Least Poorwill and Bicknell's Thrush). This area was also perhaps the last bastion of the Critically Endangered Ridgway's Hawk in the south-west of the country.

# The North

There are relatively few very productive birding sites in the north of the country, but if you do enter the country via the charter flight airport of Puerto Plata then you can easily access Monte Cristí National Park, in the extreme north-west of the country, and can also more easily visit the Samaná Peninsula. The largest town on the peninsula, Santa Bárbara de Samaná, is the centre for the Dominican Republic whale-watching 'industry'. Landlubbers can seawatch for whales from any suitable area beyond Los Cacaos, east of Santa Bárbara, where White-necked Crow is also occasionally seen. The period January to March is probably best for whales, during which season Humpback Whales *Megaptera novaeangliae* may be found offshore, especially around the so-called 'Silver Bank' (Banco de la Plata). A variety of other cetaceans are also possible.

## Monte Cristí National Park *(See Site 9 on Dominican Republic map)*

This site is principally of interest for its large numbers of waterbirds, although it is not necessarily an easy area to work, and is also the gateway to the small offshore islands, the Cayos Siete Hermanos, which harbour colonies of breeding seabirds in summer.

***Location*** If staying at Puerto Plata, take Highway 5 west out of town and then follow the road south until you reach the intersection with Highway 1, from where you should turn right and continue the approximately 90km to the town of Monte Cristí. To reach the main lakes, Laguna Salinas and Laguna Saladilla, head south from the town on Highway 45, in the direction of Manzanillo and Dajabón. Just after the Nueva Judea sign on the left and on reaching the national park station on the same side of the road, turn right onto a dirt road and continue to the end, from where you can view the lagoon. To reach Laguna Saladilla, return to Highway 45 and proceed south again, through Los Conucos until you reach the right-hand turn to Manzanillo (called Pepillo Salcedo on some maps). Shortly afterwards, you should take the left turn on a dirt road that leads to an aqueduct, from where the east shoreline of the lagoon soon comes into view. Reeds obscure most of the lake but you may be able to hire a boat to explore the area. Return to the Manzanillo road and continue west to the town; from the Puerto Cristal restaurant at the mouth of the river here, you can search the mangrove for species such as Clapper Rail. Head for the pier at the north end of Manzanillo and then take the track on the right to reach the area.

***Accommodation*** Most visitors are likely to day-trip the area from one of the resort hotels around Puerto Plata, but bear in mind that West Indian Whistling Duck is most easily seen around dusk, which will mean driving back to your hotel after dark. A range of hotel options is available in Monte Cristí.

***Strategy*** One of the best areas for the globally threatened West Indian Whistling Duck is Laguna Saladilla, especially at dusk or early in the morning. Areas of shoreline north of Monte Cristí can be productive for a variety of waterbirds, including Wilson's Plover, Yellow-crowned Night Heron, many other herons, gulls and terns, and Clapper Rail. In winter very large numbers of immigrant waterfowl were formerly present on a regular basis in this region, but numbers are currently much lower. Nonetheless, this is still one of the most productive areas of Hispaniola for such species. Plain Pigeon can be found in any suitable patch of habitat in the national park, and scrubby areas can hold typical species such as Vervain Hummingbird, Loggerhead Kingbird, Black-crowned Palm Tanager, Yellow-faced Grassquit and Hispaniolan Oriole.

The offshore Cayos Siete Hermanos are worth visiting during May to

August, during which period Brown Noddy (50–380 birds) and Bridled (1,000–2,500 birds) and Sooty Terns nest on the islands. To visit these islands it is essential to go to the national park headquarters just north of the town centre, on the road towards El Morro, from where a boatman to make the trip out to the islands can also be engaged.

***Birds***

The following are some of the species possible in this area, although you may need some time to find a significant proportion of these: Least Grebe, Reddish Egret, Yellow-crowned Night Heron, White Ibis, Roseate Spoonbill, American Flamingo, West Indian Whistling Duck, Red-tailed Hawk, Clapper Rail, Wilson's Plover, Least, Bridled and Sooty Terns, Brown Noddy (these last three best looked for in June or July on Cayos Monte Chico and Ratas), Plain Pigeon (rare), White-winged Dove, Hispaniolan Parakeet, Mangrove Cuckoo, Antillean Nighthawk (summer), Antillean Palm Swift, Antillean Mango, Vervain Hummingbird, Broad-billed Tody, Hispaniolan Woodpecker, Grey and Loggerhead Kingbirds, Stolid Flycatcher, Caribbean Martin, Cave Swallow, Palmchat, Black-whiskered Vireo, Yellow Warbler, Black-crowned Palm Tanager, Yellow-faced Grassquit, Greater Antillean Grackle, Hispaniolan Oriole, Nutmeg Mannikin and Village Weaver (the last two introduced).

**OTHER SITES**

Plain Pigeon can be found at **Laguna Gri-Gri**, just south-west of the town of Río San Juan on the north coast between Sosúa and Cabrera, accessed off Highway 5. Check the mangroves around the lake.

Hispaniolan Parakeet can sometimes be found at the eastern end of the holiday resort of **Sosúa**, in the strip of woodland strip between Playa Escondida and the hotel of the same name.

If you are on a largely non-birding holiday, based at one of the north-coast resorts, then the **El Chocó Reserve**, signed 500m east of Cabarete on Highway 5 just south-east of Sosúa, is a good option for a morning's birding. Drive inland through the village of Cabarete to the reserve entrance, and then walk the rough track to the T-junction, where it is best to turn right, as the left turn goes to the tourist caves (for more information see www.hispaniola.com/cabaretecaves). Birds possible here include: Sharp-shinned Hawk, White-crowned and Plain Pigeons, Mangrove Cuckoo, Palmchat, Antillean Palm Swift, Vervain Hummingbird, Antillean Mango, Hispaniolan Emerald, Hispaniolan Woodpecker, Antillean Piculet, Hispaniolan Pewee, Black-whiskered Vireo, Black crowned Palm Tanager, and Greater Antillean Bullfinch.

# PUERTO RICO



Discovered by Christopher Columbus during his second transatlantic voyage, in 1493, Puerto Rico is the fourth-largest island in the Greater Antilles and a popular birding area, especially among North American birders. This US Commonwealth territory of 880,190ha is the main embarkation port for many Caribbean cruises, as well as a favourite vacation area in its own right. Except for its Hispanic ancestry (the island became an American colony at the end of the Spanish-American war in 1898), the island could easily be mistaken for Miami or San Diego. It has all of the modern conveniences and stores associated with the USA, including Target, Venture, K Mart, McDonalds, Kentucky Fried Chicken and Pizza Hut. The general ambience and environment is quite unlike any other of the major islands of the Greater Antilles. Many Puerto Ricans speak English, but learning a few Spanish phrases will prove helpful and will be appreciated by the local people (as well as standing you in good stead in Cuba and Hispaniola).

The heart of Puerto Rico is San Juan, the island's capital, where approximately one third of the population lives. Within the city, a seven-block area known as Old San Juan (Viejo San Juan) contains a sprinkling of 16th and 17th century landmarks, as well as sporting many cafes, museums, and galleries. Casa Blanca, constructed in 1523, housed the island's first Spanish governor, Ponce de Leon, who first visited Puerto Rico in 1508 and noticed the gold trinkets of the native population. La Fortaleza, built in 1540, is the New World's longest continuously occupied executive mansion. The waterfront battlements of El Morro held off foreign advances, including one by the famous English pirate cum navigator, Sir Francis Drake. There were also several French attempts to overrun the island. The Cathedral de San Juan, one of the earliest churches in the Western Hemisphere, and the Museo de Pablo Casals, which honours the famous cellist, are also in the neighbourhood.

San Juan is a large city and crime is as much of a problem as in any other large city of the world (but not unusually so). However, like almost anywhere, as soon you get away from the metropolis you will encounter friendly people, and great birding.

Natural attractions throughout the countryside include the Caribbean National Forest (El Yunque), forest preserves for hiking, caves, coffee plantations, old sugar mills, and hundreds of beaches. Phosphorescent bays lie off the south-west coast, while the Arecibo Observatory, on the north coast, west of the capital, hosts the largest radio telescope in the world. Puerto Rico is almost rectangular in shape; it is the easternmost island of the Greater Antilles and has the highest population density in the Caribbean. The country is mountainous, except for the coastal lowlands and a belt of karst limestone, which extends from Aguadilla in the west to Loiza, east of San Juan. Approximately 7% of the island's land area is nominally protected for nature conservation (or 66,800ha) and forest in various stages of regeneration covers 42% of Puerto Rico.

As with all of the islands in this book, the main attraction for birders on Puerto Rico is the endemics, which number about 18 species (taxonomy-dependent). All, bar the parrot, might be found in a 24-hour period, by the dedicated and determined, although of course most will plan on visiting the island for at least two or three days, if not a week.

Birders not planning to visit the Lesser Antilles in the near future will also want to put in some time looking for the several hummingbirds that reach their northernmost outpost on the island, as well as the Bridled Quail-Dove. Flights between Puerto Rico and Hispaniola are frequent, meaning that the two islands might easily be combined, if you have ten days or so to spare. Birding in Puerto Rico is easy, with the roads being good to excellent and nowhere very far away, although traffic can be very heavy on some of the major routes at peak times (this is a much greater problem than on any other of the Greater Antilles). Car rental is reasonable, being comparable to Florida and Hawaii, and all of the continental US rental companies are represented here. If you arrive by plane, there is a mid-priced hotel in the airport itself, appropriately named the International Airport Hotel. If you arrive late at night, you can save the cost of a rental car by staying there, rather than going into San Juan or Fajardo. After a good sleep, you can pick up your rental car early in the morning. If you are entering Puerto Rico to join a cruise, this hotel would also be a good place to stay, as you could return to the airport, drop off your car, and then take a shuttle to the pier. There are also budget hotels in nearby Miramar, including the reasonably well recommended Hotel Toro ($45 double), telephone 809-725-5150 or 809-725-2647.

*Map of Puerto Rico*

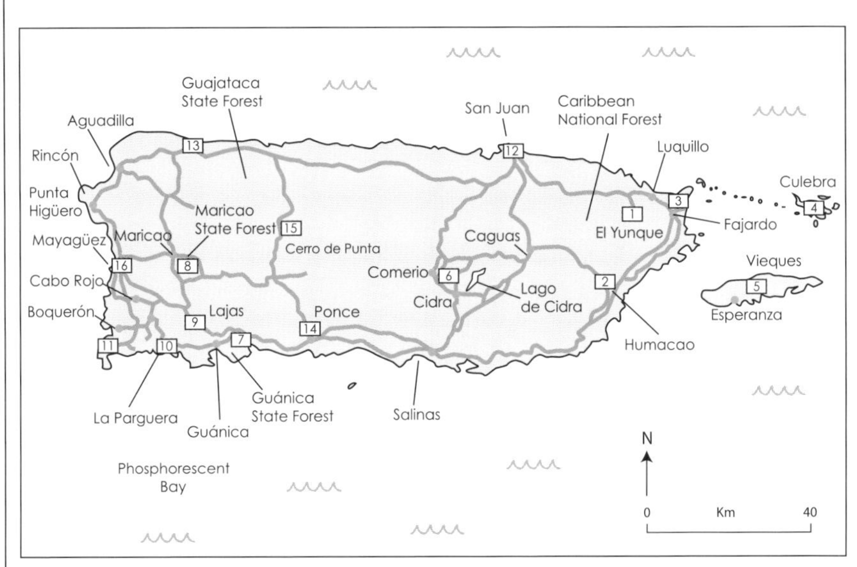


*Strategy*

If you decide to arrive early for a cruise or another Caribbean location to bird, we would recommend setting aside four days to see the endemics, although you can make a good dent in the endemic list even if you just manage a full day. Puerto Rico is easily visited during a stopover while in transit to other Caribbean destinations. If you can spend an evening and a day in Puerto Rico, you can probably see all the endemics with the exception of the parrot. If you arrive in the afternoon get a rental car and drive south to Guánica. If the reserve is open, look for Puerto Rican Lizard Cuckoo, Puerto Rican Tody, Adelaide's Warbler, Pearly-eyed Thrasher and Puerto Rican Bullfinch. If you arrive at dusk or later, follow the directions in the site information for the owl and the nightjar. Stay the night near Lajas and see the Yellow-shouldered Blackbird at dawn. Then drive to Sabina Grande and take highway PR120 to Maricao, stopping along the way for Puerto Rican Tanager, Puerto Rican Vireo, Elfin Woods Warbler and Puerto Rican Pewee. Along the way you should find Puerto Rican Woodpecker, Puerto Rican Emerald, Green Mango, and the others. Return to San Juan and, if time permits, go to El Yunque or Río Abajo Forest to try for the parrot.

## El Yunque (the Caribbean National Forest)

*(See Site 1 on main Puerto Rico map)*

This 11,200ha area in the Sierra de Loquillo is administered by the US Department of Agriculture's Forest Service, and is the only tropical forest in their system of preserves. The area was first set aside for protection, in 1876, by the Spanish, thereby making it one of the oldest reserves in the Western Hemisphere. More recently, the area was declared a Biosphere Reserve by UNESCO and an Important Bird Area by BirdLife International. There are three main peaks within the National Forest, all of them just over 1,000m in altitude, and the highest being El Toro, at 1,074m, in the south-west. It is the wettest national forest – up to 5,000mm of rain fall on the highest peaks per annum, while the mind-boggling total of approximately one billion gallons of water may fall per year. The combination of this heavy rainfall and the tropical climate has created a dense evergreen forest encompassing 240 native tree species (23 of them endemic to this forest), along with many vines, epiphytes, giant ferns and mosses. Twenty-six species of plants are endemic. True rainforest (Tubonuco) occurs on the lower slopes below 600m. There is a trail map online (www.fs.fed.us/r8/caribbean/recreation/recreation_hking.shtml).

***Location*** To reach El Yunque from the international airport, take the southbound lanes on highway PR26 and follow the signs to Carolina. On reaching Carolina, turn east on highway PR3 and follow this road beyond the towns of Canóvanas and Rio Grande, until you reach the turn-off to El Yunque, which is highway PR191 (note that the sign is easy to miss). Turn right on PR191, then immediately right again at a T-junction with a garage, and then next left in a small village (known as Mameyes or Palmer). The road winds uphill, and eventually you will pass through another village (Barcelona), before reaching the entrance to the park. PR191 continues and winds through the park, before terminating at a locked gate before the PR9938 intersection.

***Accommodation*** Fajardo is arguably the best area in which to stay to access the Caribbean National Forest, as well as Humacao National Wildlife Refuge and Culebra Island. We would recommend the Parador Familia, but the following have also been recommended. The more expensive Fajardo Inn (signed off the PR3 beside the Shell station at the entrance to the PR194; turn east here and continue on this road until you reach the junction with the PR195, where you should turn right towards the ferry terminal, and continue until you see the hotel building on a small hill to the left up a side street), 52 Parcelas Beltrán, Puerto Real, PR 00740 (telephone: 787-860-6000; website: www.fajardoinn.com), is a known site for Antillean Crested Hummingbird and Green-throated Carib in the grounds. Also here is Pearly-eyed Thrasher. Particularly check the flowering trees close the hotel reception for the hummingbirds. Prices for a double room start at $130. Other places to stay include the Anchor's Inn (telephone: 787-863-7200), Las Delicias Hotel (telephone: 787-863-1818), the Scenic Inn (telephone: 787-860-6000) and the Ceiba Country Inn (telephone: 787-885-0471; website: www.lanierbb.com/inns/bb6488.html), which are all reasonably priced ($60–70 a double) and clean! The final option currently has a pair of Puerto Rican Screech Owls on site. Further west, along highway PR3, near Rio Grande, 5.7km east of the PR186 turn is the Motel El Rio (PR967, Barrio Jiménez; telephone 887-5486), which is a 'love hotel', with mirrors covering the walls and ceilings, and a double bed with a rubber sheet. The upside is that the 'accommodation' is far cheaper and fine for the low-budget birder whose pride is not at issue. There is also the Hotel Yunque Mar (telephone: 787-889-5555; website: www.yunquemar.com), which is well signed off highway PR3 at Fortuna, is reasonably priced and puts you within a 15-minute drive of the entrance to the National Forest. Some innovative birders have even slept in the unlocked female park employees' dressing room at the Palo Colorado picnic area in El Yunque, but this seems even more desperate than using the motel! There are many restaurants in Luquillo or Fajardo, both local and franchised, e.g. McDonalds, Subway, KFC, Pizza Hut, K Mart, Walgreens, and many other grocery and department stores as well.

***Strategy***

Bird the following areas. Bear in mind that the area can be very busy with tourists and sightseers at weekends. A formerly regular site for the Critically Endangered Puerto Rican Parrot is the Rio Espiritu Santo lookout on the PR186, which is a few hundred metres south of where the road crosses a river, in the extreme north-west of the National Forest. This used to be a regular site for the parrot, but they have become very erratic here in recent years and the lookout itself is now blocked off by bamboo and a huge *Cecropia* tree; nonetheless, observers have occasionally been fortunate here even in the last couple of years.

**A:** About 0.4km before the entrance to the National Forest there is a deserted house on the right-hand side of the road. Nearby, you may see many of the open-country birds, and this is also a good site for Puerto Rican Lizard Cuckoo, while Puerto Rican Screech Owls can be found nearby. The cuckoo is best seen in the brushy disturbed areas between Mameyes, at the bottom of the hill, and the park entrance (km8.1 from highway PR3). A tape will prove very useful in finding both the lizard cuckoo and the screech owl. The La Coca Falls immediately inside the entrance gate can be a productive area to find either of the endemic Puerto Rican hummingbirds.

**B:** On entering the park, the next area of interest is the Yokahu Lookout Tower (at km8.9 from highway PR3), on the left-hand side of

*Map of El Yunque plus inset map*

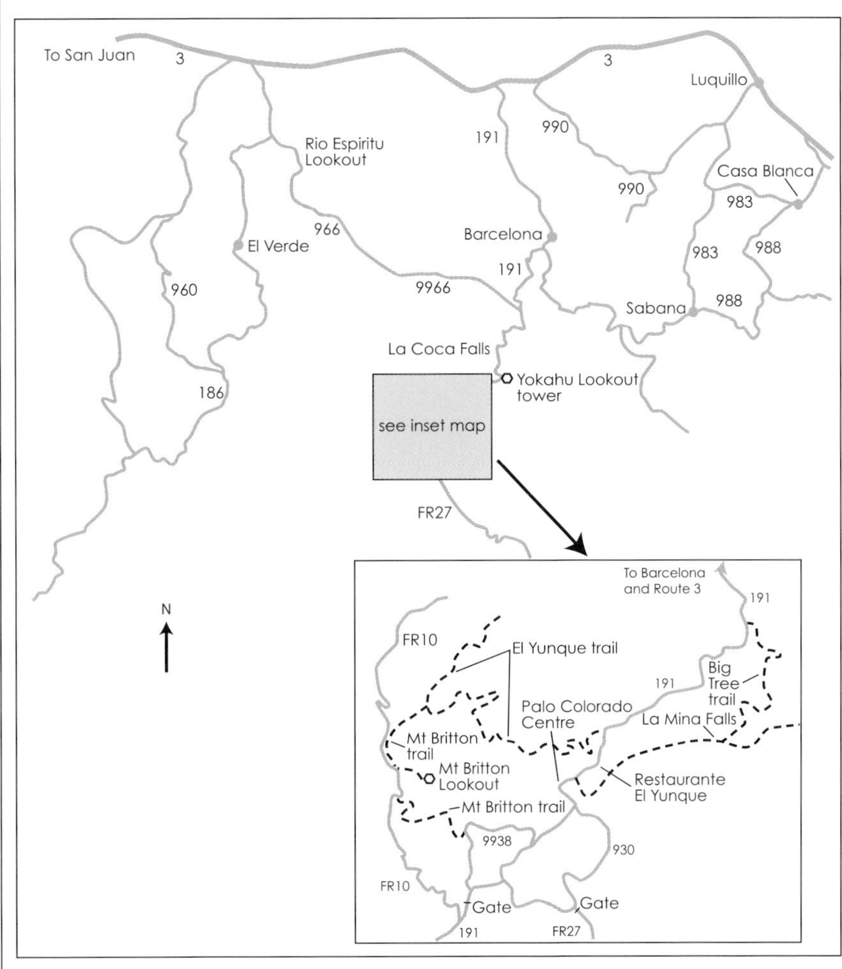


the road as you climb the hill. If you scan the valley from here, you should see some tree platforms. These could be occupied by researchers, who are observing the movements of Puerto Rican Parrots. Do not expect to see parrots in this area (as the wild population in El Yunque now numbers a mere 18 individuals), but birding can be good around the tower and on the nearby trail. In particular, this is a great location for Black Swift (present from mid April).

**C:** Further up the road, again on the left-hand side, a stop at the Big Tree Trail (km10.2) may prove productive. At dusk, Puerto Rican Screech Owl is sometimes found around the parking lot, while during the day Green Mango (which is relatively rare in the eastern half of the island) has occurred here, especially when suitable flowers are abundant.

**D:** To bird the El Yunque Recreational Area continue along PR191 until you reach the Sierra Palm Information Center, which will again be on your left (km11.6). Restrooms and parking are available. Walk back down the road and on the right look for an entrance that is chained-off. If you park nearby and walk down this overgrown road you will reach the ruins of the 'El Yunque' Restaurant, which was destroyed by

Hurricane Hugo in 1989. This is still a good birding area, and the gorge of the Río de la Mina below you is one of the major flyways of the remaining Puerto Rican Parrots (Critically Endangered). You will see some more of the observation platforms across the valley from the veranda. Scaly-naped Pigeon and Puerto Rican Woodpecker are easily seen in the canopy from here, and this also constitutes a good vantage point to find the Puerto Rican subspecies of Broad-winged Hawk (*brunnescens*).

**E:** Continuing still further along the PR191, on the left you will see the Palo Colorado Center (km11.8) and the 'El Yunque' trailhead. There is a previous trailhead at Sierra Palm, but the present route is recommended, as you have a fair but unsurprisingly not great chance of seeing parrots flying down the valley. In addition, you will have an excellent chance of seeing many of the more regularly observed endemics, including Puerto Rican Emerald, Puerto Rican Woodpecker, Puerto Rican Tody, Puerto Rican Tanager, Puerto Rican Spindalis and Puerto Rican Bullfinch.

**F:** The Mount Britton Trail, which levels off at 780m, branches off the El Yunque trail to the left at the ridgetop. If you follow this trail you will enter elfin forest, where the trees are stunted and moss-covered. This is the habitat of Elfin Woods Warbler, a bird that remarkably was only discovered as recently in 1971. Be prepared to spend some quality time searching, as this warbler does not necessarily respond to tape readily; it 'slithers' and creeps its way through the branches and along the trunks of trees, often making the bird difficult to observe very well. This section of the Mount Britton trail connects the El Yunque trail to a paved road, the FR10. If you continue on the El Yunque or Los Picachos trails you can reach the highest peaks.

**G:** This section of the Mount Britton trail, and the elfin forest, terminates at the FR10 road, which has little traffic, other than Forest Service vehicles, since it runs into the PR191, which is closed off below. If you walk the road, it is possible to get good looks at many of the endemics, and you will also pass the captive-breeding area of the Puerto Rican Parrots on the left. The compound is surrounded by a chain fence, but you might catch a glimpse of the captive birds as you pass. Continue along the road to the chain across the PR191 (km13 from highway PR3), walk around the barrier and return downhill to the Palo Colorado Center.

**H:** Another possibility for accessing warbler habitat more quickly is by driving the PR191 to the barrier, and turning right on the PR9938 to reach the trailhead at the other end of the Mount Britton trail. The hike is less than 20 minutes to the intersection of the Mount Britton trail and the FR10, where the habitat begins. Note that route PR9938 loops back to the PR191, but has been closed at the Mount Britton trailhead.

***Birds***

The following endemics, near-endemics and other species of interest are possible at El Yunque: Sharp-shinned, Red-tailed and Broad-winged Hawks, Scaly-naped Pigeon, Zenaida and White-winged Doves, Ruddy

Quail-Dove, Puerto Rican Parrot, Puerto Rican Lizard Cuckoo, Puerto Rican Emerald, Green Mango, Black Swift, Puerto Rican Screech Owl, Antillean Nighthawk, Puerto Rican Tody, Puerto Rican Woodpecker, Puerto Rican Pewee, Grey Kingbird, Red-legged Thrush, Caribbean Martin, Cave Swallow, Black-whiskered Vireo, Pearly-eyed Thrasher, Elfin Woods Warbler, Antillean Euphonia, Puerto Rican Tanager, Puerto Rican Spindalis, Puerto Rican Bullfinch, Yellow-faced and Black-faced Grassquits, and Greater Antillean Grackle.

## Humacao National Wildlife Refuge *(See Site 2 on main Puerto Rico map)*

This is a good area for Caribbean waterfowl, with the principal highlight being the globally threatened West Indian Whistling Duck,

*Map of Humacao*

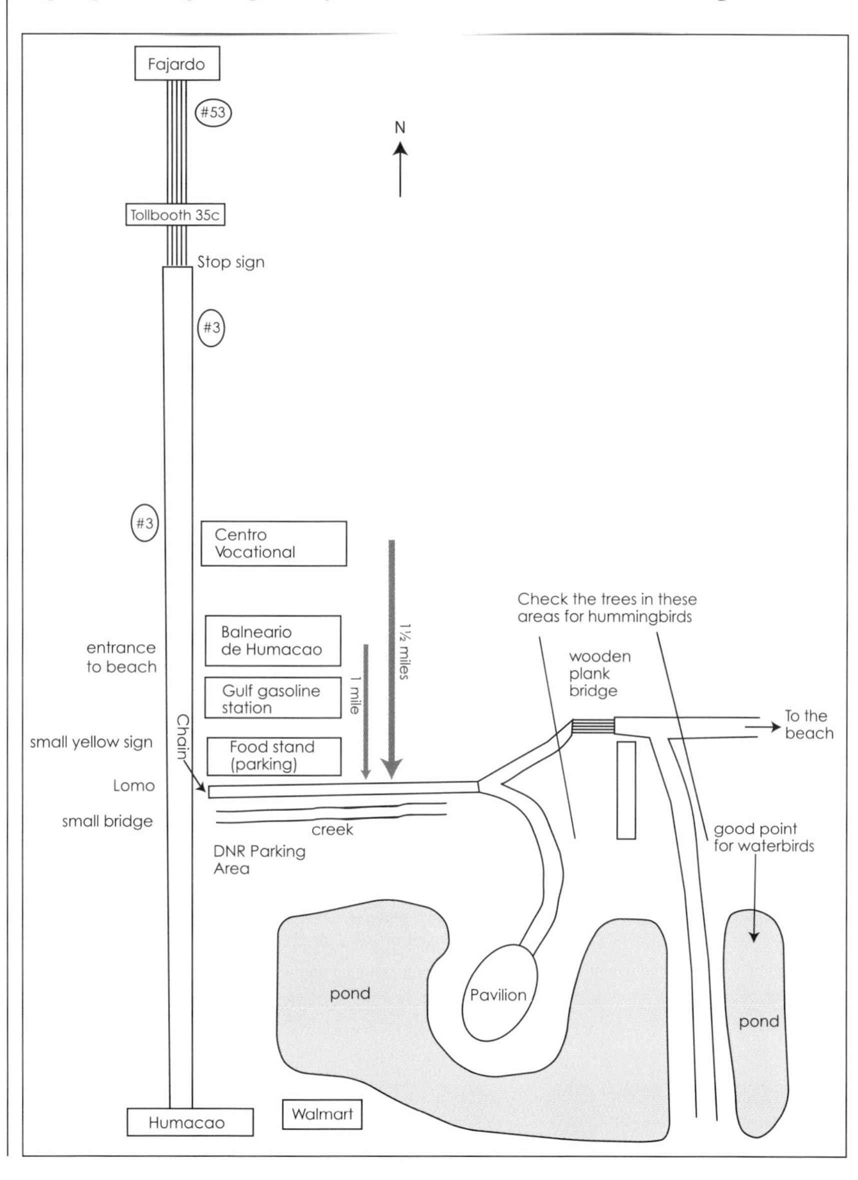

which is best searched for towards the end of the day. The refuge covers almost 1,400ha and its beaches are breeding sites for a trio of rare oceanic turtles, Leatherback *Dermochelys coriacea*, Hawksbill *Eretmochelys imbricata* and Loggerhead Turtles *Caretta caretta*.

***Location*** To reach this site from Fajardo, take highway PR53 (a toll road) south and continue as far as the turn-off to the town of Humacao and Punta Santiago. Exit here and then turn right towards Punta Santiago, continue, passing the turn to Pasto Viejo on your left, until you see a lake on your right. The entrance to the refuge is well signed soon after on the right (although it was closed for renovation in spring 2009). Alternatively, take the exit about 6km beyond the toll booth south of Fajardo, then turn left to cross below the highway and after that turn right onto highway PR3 towards Humacao. Remain on this road the 14km to the refuge. The road has many twists and turns; eventually the sea will appear in sight on your left. Look for the Centro Vocational School, again on the left, which is approximately 2.4km from the trail into the refuge. Another 0.8km past the school is the road to Balneario de Humacao. After you pass this road (to the beach), you will see a Gulf service station on the left. Slow down and park on the wide shoulder. (There may be a abandoned food stand here.) The refuge entrance is well signed on the right-hand side, but if you happen to miss this, you will pass a sign, 'Lomo', at a small bridge; just beyond the bridge is a sign for the refuge on the left. There is a parking lot here. If you arrive here by mistake, just turn around and return over the hump in the road (the 'lomo').

Coming from the west end of the island, drive east on the PR2 to the junction with the PR52, and then continue east and north as far as the PR30, at Caguas, which you follow east and south. Near the town of Humacao, this road splits and to reach the refuge you need to take the PR3, which passes under the PR53 and then continues to the entrance to the refuge, after about 5km, on the right (see also above).

***Accommodation*** We recommend that if you visit the refuge in the evening, to look for West Indian Whistling Duck, that you backtrack to Fajardo (see options under the previous site) or go on beyond Humacao. You will not find a reasonable place to stay in Humacao, it being a very pricey resort. Instead, we would recommend taking highway PR30 to Caguas (many hotels); from here you will be in position to look for Plain Pigeon in the morning.

***Strategy*** The refuge is only open for limited hours on weekdays (closed weekends). The refuge headquarters compound is open 07.30–15.30, but the back pond can be reached by taking the trail parallel to the canal immediately east of the entrance. After a short distance this emerges at an old bridge. Do not turn right towards the pavilion, but instead go left for 30m across a wooden plank bridge. At the next fork in the trail, take the right-hand path for about 100m, whereupon you will see an impounded marshy area on the left. There is a second lagoon further on, at which the trail terminates in fence. The same ponds can also be reached by walking towards the ocean from the refuge headquarters, then crossing two wooden bridges and taking the right-hand trail (to Laguna Palmas) rather than following the sign to the beach ('playa'). This spot is better in the evening for West Indian Whistling Duck (at the edge of the mangrove), but Caribbean Coot, White-cheeked Pintail and Masked Duck should be visible at virtually any time of day. This is also a good area for Antillean Crested Hummingbird (in the trees around the overlook) and Green-throated Carib is also found here but is decidedly less common than the former species.

***Birds*** The following species may be found at the refuge: Pied-billed Grebe, Least Bittern, Tricoloured and Yellow-crowned Night Herons, West Indian Whistling Duck, White-cheeked Pintail, Ruddy Duck, Northern Bobwhite, Sora, Yellow-breasted Crake (rarely seen), Purple Gallinule, American and Caribbean Coots, Wilson's Snipe, Zenaida and White-winged Doves, Puerto Rican and Mangrove Cuckoos, Puerto Rican Screech Owl, Antillean Crested Hummingbird, Green-throated Carib (uncommon), Puerto Rican Tody, Puerto Rican Woodpecker, Puerto Rican Flycatcher, Grey Kingbird, Caribbean Martin, Cave Swallow, Pearly-eyed Thrasher, Adelaide's Warbler, Nutmeg and Bronze Mannikins, Orange-cheeked Waxbill and Orange Bishop (all of the last four introduced), Puerto Rican Spindalis, Yellow-faced and Black-faced Grassquits, and Greater Antillean Grackle.

## Cabezas de San Juan *(See Site 3 on main Puerto Rico map)*

This area, close to Fajardo, is located in Las Croabas, on highway PR987, near the Parador La Familia, and is a 126ha privately owned nature preserve on the waterfront; it is managed by the Conservation Trust of Puerto Rico. It is a relatively undisturbed area of coastal scrub, and admission is by reservation only, between Wednesday and Saturday (telephone: 787-722-5882 to arrange access). The public beach borders the fenced reserve, but the fence peters out after 0.4km; many trails connect the beach with the main preserve road. Flowering mimosa trees along this road attract Antillean Crested Hummingbird and Green-throated Carib (specifically check the trees at 18°21'910"N 65°37'997"W), both of which species may also be found in the vicinity of the Fajardo Inn (see the Accommodation section of El Yunque). There are many trails, mangroves and lagoons to explore, and many of the regular endemics can also be observed here. Other species that are possible include Orange-fronted Parakeet (introduced), Antillean Mango, Red-legged Thrush, Scaly-breasted Munia (introduced) and Yellow-faced and Black-faced Grassquits. The globally threatened Yellow-shouldered Blackbird has declined dramatically in this region, but is still present in small numbers in the nearby Ceiba State Forest, while for those whose interest extends beyond birds, West Indian Manatee *Trichechus manatus* (another threatened species) is present in the offshore waters.

# Culebra Island *(See Site 4 on main Puerto Rico map)*

***Map of Culebra Island***

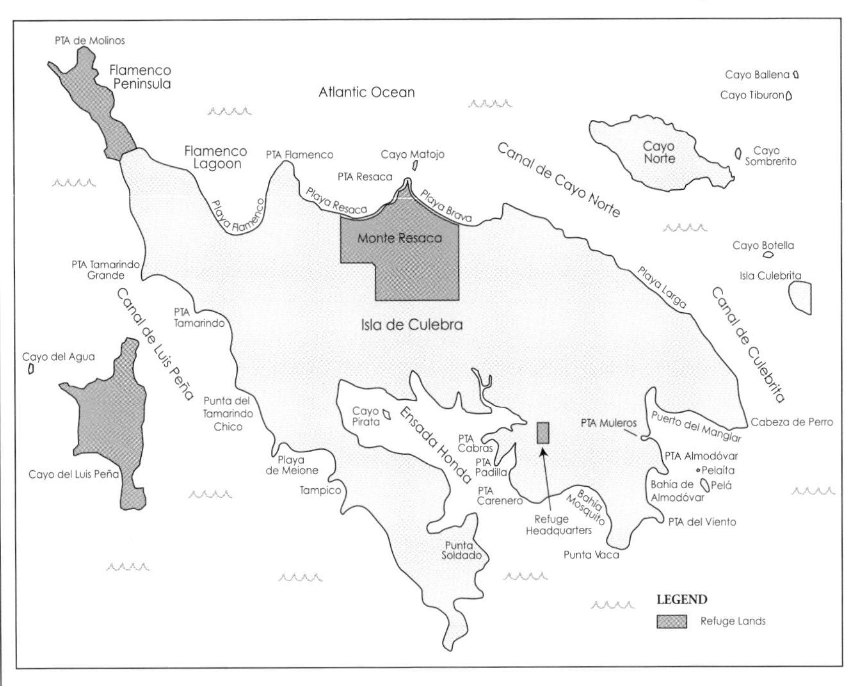


Those on a longer trip to Puerto Rico, especially in summer when the seabirds are nesting, may consider visiting this island. To do so, park your car at Fajardo and take the ferry to Culebra (the telephone number for the ferry terminal in Fajardo is 787-863-4560). It is possible to take a car across, but it is essential to make a reservation on the ferry to do so, and it is also important to reserve your accommodation in advance if planning to visit the island in the high season. When you arrive on the island you will feel like you have returned to the 1960s. A laid-back community awaits you here. You can make arrangements locally to dive or snorkel (described as spectacular for those interested in tropical fish), as well as exploring the tiny offshore cays, swim and sunbathe at Flamenco Lagoon, observe fish-eating bats from the pier at night, or just 'chill out'. There are very few cars on the island, so if you wish to go somewhere all you need to do is wave and someone will pick you up. A nice place to stay is Posada La Hamaca (telephone: 787-742-3516) and also recommended is the Club Seaborne guesthouse, which has feeders in the garden (Green-throated Carib, Pearly-eyed Thrasher and the introduced Venezuelan Troupial have all been recorded here). The Palmetto guesthouse (telephone: 787-742-0257 or 235-6736; e-mail: palmettoculebra@yahoo.com; website: www.palmettoculebra.com) offers free transfers from the ferry port.

Culebra is a small island of 2,800ha, sections of which have been designated a National Wildlife Refuge since 1909. This refuge comprises 590ha, contains 23 islands and rocks, and four tracts on the island. Green, Leatherback and Atlantic Loggerhead Turtles nest here, and their numbers are monitored by the CORALations organisation, which is also undertaking coral restoration activities. Mount Resaca is on the north coast; it protects a large block of dry subtropical forest, which harbours bird species such as Puerto Rican Flycatcher and Adelaide's Warbler. The Flamenco peninsula on the north-west coast is a refuge for nesting Sooty Terns (the population has been estimated at 105,000 birds). Nearby cays support large numbers of breeding Laughing Gulls, Bridled, Sooty, Roseate, Royal and Sandwich Terns, and Brown Noddy (up to 1,860 birds), as well as small colonies of White-tailed and Red-billed Tropicbirds. In addition, small groups of Masked and Red-footed Boobies nest here, along with a large colony of Brown Boobies. Brown Pelicans frequent the mangroves. For more information and a bird list, call or write to the Refuge Manager, Culebra NWR, PO Box 190, Culebra PR 00775 (telephone: 787-742-0115).

## Vieques Island *(See Site 5 on main Puerto Rico map)*

Some of the ferries from Fajardo en route to Culebra stop at Isla de Vieques, and there is also a direct ferry to this 13,500ha island of rolling hills, which lies 13km off the Puerto Rican mainland. Beaches are the main attraction here, but several of the Puerto Rican landbird endemics can be found, as well as Brown Booby, Red-billed Tropicbird and Magnificent Frigatebird. More than 100 species have been recorded (for a complete island checklist, see www.bigpockets.com/roost/caribbean/vieques_checklist.html).

Most importantly any accessible area of forest should be searched for Bridled Quail-Dove, especially the Vieques National Wildlife Reserve (VNWR), where both Ruddy and Key West Quail-Doves are also present. Key West Quail-Dove occurs in the palmetto dry forest on the southern limestone coast Puerto Ferro and Puerto Mosquito (VNWR) and Puerto Mosquito and Laguna Sombe, but the former area is currently closed until a toxic waste cleanup is complete. The most reliable place is at the end of the road from Laguna Sombe to Puerto Mosquito, at the parking lot for the electric boat ride on Puerto Mosquito Lagoon, and the best time is first light (especially in March–June). Walk back along the road and take the first left on a dirt road that ascends the cliffs. Bridled Quail-Dove prefers higher elevation moist forest, such as that found about halfway to the top of Mount Pirata (VNWR), which forms the highest point on the island, at 300m. Walk up from the locked gate at the base of the mountain and especially check the area just downhill of where the power line swath heads steeply uphill on the left, but it is rare to catch more than a brief view as a bird flies across the road. Ruddy Quail-Dove seems to be the most widespread *Geotrygon* here, preferring hilly, densely wooded areas in the west of the island, and again the best place to hear the species is at the locked gate on the road up Mount Pirata (VNWR).

Significant numbers of White-cheeked Pintail move between Vieques, Culebra and the eastern Puerto Rican mainland. Boca Quebrada at the western end is being cleaned up and may soon be accessible to birdwatchers, while another possible spot is just west of Puerto

Mosquito. Take the dirt road off Route 997 to Puerto Mosquito, where the kayaks put in for the Bio Bay tour. An old road that can be inundated and very muddy leads south of the parking area, and after about 400m takes a sharp turn to the right, where it skirts a lagoon and eventually ends on the east side of Laguna Sombe. Pintails can usually be seen in these well-vegetated lagoons, while Clapper Rail, many shorebirds and Blue-winged Teal can also be found, and Black Rail and Chuck-will's-widow have been recorded. Adelaide's Warbler, Puerto Rican Flycatcher and Loggerhead Kingbird inhabit the scrub. Other species of interest found on the island include Puerto Rican Screech Owl, Puerto Rican Woodpecker and Antillean Crested Hummingbird, while West Indian Manatee, and Leatherback, Hawksbill and Loggerhead Turtles all occur offshore.

New and unusual sightings and/or comments on the seasonal status of the island's birdlife should be sent to the Vieques Conservation and Historic Trust, Calle Flamboyán 138, Vieques, PR 00765, (telephone: 787-741-8850, e-mail: vcht@coqui.net). The trust may also be able to advise on additional birding opportunities on the island. More than two-thirds of the island was formerly a US Navy base, including a bombing range, but as a result of prolonged protests the navy pulled out in 2003 and most of the island is now a designated US Fish & Wildlife Service reserve. General information on transportation, accommodation and other facilities can be obtained by visiting: www.isla-vieques.com/puerto-rico/introduction.htm.

## Caguas *(See Site 6 on main Puerto Rico map)*

On Puerto Rico, the globally threatened Plain Pigeon is represented by an endemic subspecies and is restricted to that area of the island bounded in the east by San Lorenzo, in the north by Caguas, in the west by Comerío and in the south by Aibonito and Cayey. One readily accessible site for the species is near Comerío. Heading south on PR52, take exit 19 at Caguas onto PR156 heading west through the town of Aguas Buenas, always following the main road. After Aguas Buenas, PR156 becomes narrower and more winding, and eventually you pass the turn onto the PR791 on the left (south). Shortly afterwards turn left onto the PR172 and, at km1.5, look for the entrance to a baseball field next to the Escuela Superior Sabana. If you stand at first base and look back to the trees around the entry gate, you may see some pigeons perched, and if you spend some time here you will surely find the species, although early morning or late afternoon are certainly the best times of day. There is another site at km0.4, where Plain Pigeons roost in the mango trees either side of a small bar, and the species has also been found nesting in this area (the main nesting period on Puerto Rico is March–June). The current Puerto Rican population has been estimated to number anything between 700 and almost 4,000 individuals in the wild during the last decade and a half, and about 50 in captivity. By the mid 1930s the species was virtually extinct on Puerto Rico, but in 1958 it was discovered near Cidra, where there were 150 birds in 1988.

## Guánica State Forest *(See Site 7 on main Puerto Rico map)*

*Map of Guánica State Forest*

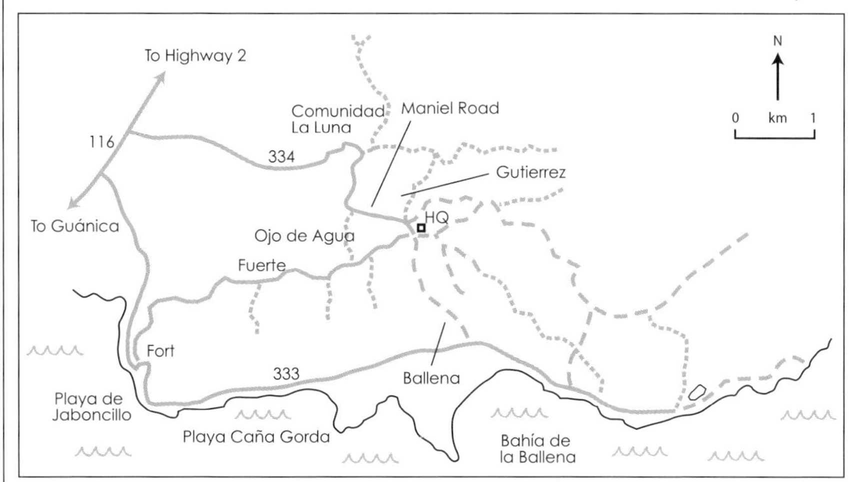


This 3,300ha area of subtropical dry forest is particularly good for Puerto Rican Lizard Cuckoo, Puerto Rican Tody, Adelaide's Warbler, and above all is the principal (and only regularly visited) site for the endangered Puerto Rican Nightjar, making it an essential destination for those intent on seeing all of the endemics. This highly restricted-range species went unrecorded between its initial discovery in 1888 and 1961. Approximately 20% of the population occurs within the protected area. The State Forest has also been designated a UNESCO Biosphere Reserve. Rainfall in the period of January to May can be as little as 30mm, and at this season there are usually very few leaves on the trees making the birds easier to see.

*Location*

The main southern freeway is the PR52, which terminates at Ponce. From the latter town get onto the PR2, which continues to Palomas, just beyond which is the turn-off, the PR116, to Guánica State Forest and the town of the same name. There are two entrances to the park, the western gate being accessed via the PR334, off the PR116 north of the town of Guánica, and the southern gate by taking the PR333, which is also signed off the PR116, past Playa Jaboncillo and around the coast. For the western entrance, leave the PR116 on the PR334 and keep left on the main road through Comunidad La Luna, winding up a slight hill, until you reach the entrance. Continue into the reserve.

*Accommodation*

Some of the best options in this region are on the PR333 coast road. At km7.2 along here, past the large Copamarina Hotel, is a stakeout for the introduced Venezuelan Troupial, which can also be seen near the beach at the terminus of the PR333, and Pearly-eyed Thrasher is common in this area too. Should the Copamarina not be to your taste, another good place to stay in this area is Mary Lee's By-the-Sea (telephone: 787-821-3600). Small rooms cost US$70–90, and larger rooms that can sleep groups cost US$110–$150. Kitchens with decks facing the sea are also available. Another beachside option is the Hotel Guánica 1929 (www.guanica1929.com). The other possibility in this general area of the island is to stay at the Hotel Villa Parador de La Parguera, at La Parguera (telephone: 787-899-7777; e-mail: pvparguera@prtc.net; website: www.villaparguera.net), which will also leave you in a good position to find the endemic and highly localised Yellow-shouldered Blackbird.

***Strategy***

For those merely intent on finding the Puerto Rican Nightjar, there is a slight problem as the park entrances are chained at night (between 17.00 and 08.30). There is a small village, Comunidad La Luna, near the first entrance, reached by following the PR334, and we understand that vandalism here is unlikely. So you can park here and walk the main road (called Maniel Road) into the park. We have sometimes taken the precaution of obtaining permission from the local police, to try and minimise any risk, but should emphasise that to date we have never had any problems. Both Puerto Rican Screech Owl and Puerto Rican Nightjar are often easily located within the first 250m inside the park. The nightjar usually sings from low perches up to 6m above the ground.

The alternative is to drive the PR333, which forms the western and southern boundary of the park, winds uphill, and is heavily trafficked. Drive slowly and listen for the guinea pig-like *weep-weep-weep* call of the nightjar (easily imitated, if you don't have a recording). As you approach the top of the hill there is an abandoned concrete 'fort' on the left. Park carefully nearby and try playback for the nightjar, which vocalises especially from November until May but, as the breeding season extends to July, the species sings all year, although with minimal activity in September and October. According to one recent study, most nests are initiated in February to early July.

You can also continue along PR333 from the fort for about a further 5km, until you come to a trail on the left (Ballena). You may walk this trail to the park headquarters, which takes only 20 minutes, along which it is possible to obtain good looks at both Puerto Rican Screech Owl and the nightjar. From the headquarters, the Fuerte trail has been recommended by several visitors; also bear in mind that mosquitoes and other biting insects can be pretty numerous in the reserve at some seasons. During the daytime, it is also worth continuing beyond the Ballena trail as far as the point where the tarmac road terminates. From here, the Ojo de Agua trail proceeds inland (another potential site for Puerto Rican Nightjar), while a lagoon a little bit further on can be productive for waders during passage periods.

***Birds***

The following endemics, near-endemics and other species of interest are possible in the State Forest: White-cheeked Pintail, Red-tailed Hawk, Wilson's Plover, Zenaida and White-winged Doves, Eurasian Collared Dove (in the local towns), Mangrove and Puerto Rican Lizard Cuckoos, Antillean Nighthawk, Puerto Rican Nightjar, Puerto Rican Screech Owl, Antillean Mango, Puerto Rican Emerald, Puerto Rican Tody, Puerto Rican Woodpecker, Caribbean Elaenia, Puerto Rican Pewee, Puerto Rican Flycatcher, Grey Kingbird, Puerto Rican and Black-whiskered Vireos, Pearly-eyed Thrasher, Caribbean Martin, Cave Swallow, Red-legged Thrush, Adelaide's Warbler, wintering North American warblers (especially Prairie Warbler), Antillean Euphonia, Puerto Rican Bullfinch, Venezuelan Troupial and Greater Antillean Grackle.

## Highway PR120 and Maricao *(See Site 8 on main Puerto Rico map)*

***Map of Maricao***

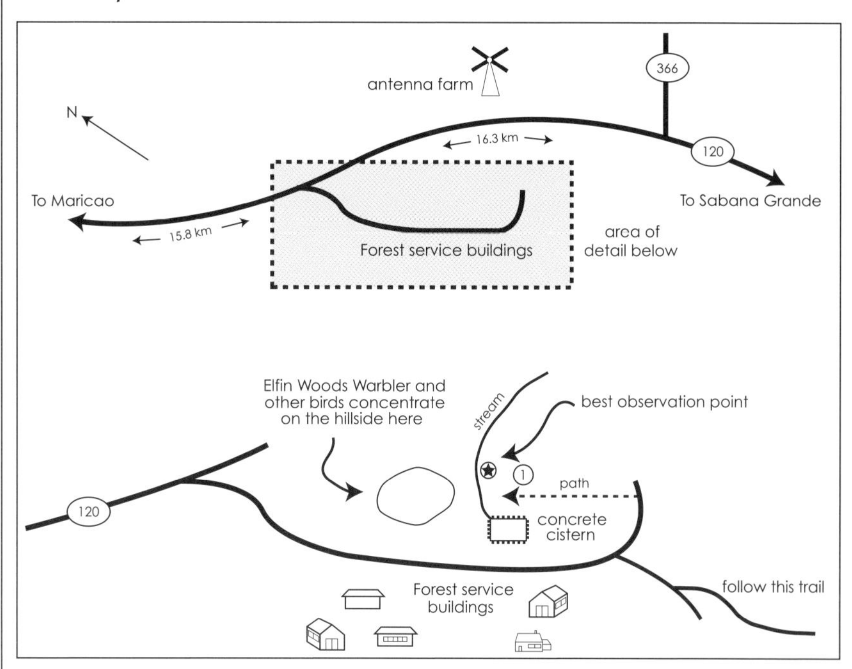


Although Elfin Woods Warbler can be observed along the Mount Britton Trail in El Yunque, the present site, which forms part of the recently designated Maricao and Susúa Important Bird Area, is by far a better and easier place to find the species. A reasonable range of other endemics is also easily found in this area in the south-west of the island.

***Location and Strategy***

From Maricao, take PR120 east and south (in the direction of Sabana Grande) 15.8km to the turn-off to the Department of Natural Resources Area, where there is a large sign 'Departmento de Recursos Naturales, Bosque de Maricao, Bienvenidos'. Drive through the gate and park by the picnic area, which is about 200m inside. Elfin Woods Warbler is seen along the stream, which flows into a concrete cistern (see Map), as well as elsewhere in the surrounding area. You can walk along this road, which has buildings and a fence to the right. As the fence ends there is a dirt road with a yellow barrier to the right. Follow this road until it forks to the left. Stay on the left fork, and bird this abandoned road for about 3km. The warbler and a range of other endemics can all be found along here. A little further along the PR120, at 16.2km is another place where the warbler is seen, and at 16.8km there is a trail where Key West Quail-Dove is sometimes seen in the early morning. Another area along PR120 is the gated dead-end road opposite the Vocational Center

(13.3km from the Hacienda Juanita), as is the dead-end road just beyond the 13km marker; look immediately to the right for a yellow-diamond warning sign, which shows a right arrow. Just behind this arrow is an overgrown entranceway to pull over and park. This area is very good: there are two overgrown trails, but the one straight ahead is best; it meanders through tall grass and ends at a stream. This was a former picnic site and you will see a pair of cement picnic benches and shelters. Further up PR120 there is a gated picnic ground. The gated road across from it is another good spot for the warbler. As you proceed further up the PR120 you will encounter La Torre de Piedra (The Stone Tower), and between km14.8 and km14.9 are several vistas for viewing the Cordillera Central and the west end of the island.

*Accommodation*

Particularly well recommended in this area is the Hacienda Juanita (telephone: 787-838-2550; website: www.haciendajuanita.com) near Maricao, which is surrounded by good secondary forest and has a couple of trails (in 2007 it cost approximately US$100 for a double room here). The hotel is on the PR105 just west of the junction with highway PR120. The grounds and garden are good for birding, with Puerto Rican Screech Owl on site, and many of the other endemics can be found in the environs.

Elfin Woods Warbler

*Birds*

Endemics, near-endemics and other species of interest found in the vicinity of the Maricao State Forest include: Red-tailed and Sharp-shinned Hawks (the endemic subspecies, *venator*, of the latter is generally very rare but probably most easily found in this area), Scaly-naped Pigeon, Zenaida and White-winged Doves, Ruddy and Key West Quail-Doves, Puerto Rican Lizard Cuckoo, Black Swift, Puerto Rican Emerald, Green Mango, Puerto Rican Screech Owl, Puerto Rican Tody, Puerto Rican Woodpecker, Puerto Rican Pewee (uncommon), Puerto Rican Flycatcher, Grey and Loggerhead Kingbirds, Puerto Rican and Black-whiskered Vireos, Red-legged Thrush, Elfin Woods Warbler, Adelaide's Warbler (uncommon), various wintering North American warblers (Black-throated Blue is particularly common), Pearly-eyed

Thrasher, Puerto Rican Tanager, Puerto Rican Spindalis, Antillean Euphonia, Puerto Rican Bullfinch, Black-faced Grassquit, Greater Antillean Grackle, Venezuelan Troupial (in gardens above Sabana Grande) and Puerto Rican Oriole.

## Susúa State Forest *(See Site 9 on main Puerto Rico map)*

This forest preserve is located just 15 minutes east of Sabana Grande and makes a great base for birding south-western Puerto Rico. There was some hurricane damage apparent in 1999, but this should have been cleaned up and repaired by this time.

*Location*

Take PR368 east from Sabana Grande for 3.4km. Turn left at the small sign for the forest (onto the PR365), and follow it to the terminus at the entrance road to the park. From here the distance is about 2.5km to the headquarters and campground.

*Accommodation*

The area is easily visited on a day trip from either Maricao or the Guánica area. It is possible to camp in the State Forest, but a permit is necessary. To arrange this, call 787-724-1374 in San Juan or 787-833-3700 in Mayagüez.

*Strategy*

Walk the various trails or check the picnic area at the entrance, where Key West Quail-Dove can sometimes be seen. A visit between November and May is likely to be most productive for Puerto Rican Nightjar, as this is the period when the species vocalises most frequently. The State Forest is one of just five known areas for this globally threatened species, and densities are similar to those encountered at the much better known site of Guánica. However, the total number of pairs at the latter site is considerably larger, as the overall area of suitable habitat is that much greater.

*Birds*

This is a particularly good area for Puerto Rican Pewee and Key West Quail-Dove, and also harbours Puerto Rican Nightjar as well. Other endemics such as Puerto Rican Lizard-Cuckoo, Adelaide's Warbler and Puerto Rican Tody are common residents, and Scaly-naped Pigeon, Puerto Rican and Black-whiskered Vireos, Puerto Rican Oriole and Pearly-eyed Thrasher are other possibilities.

## La Parguera *(See Site 10 on main Puerto Rico map)*

The 'last' of the localised endemics, the very rare Yellow-shouldered Blackbird, is best located at the Parador Villa Parguera beside an often beautifully phosphorescent sea (boat tours to see the bioluminescent plankton can be arranged at the hotel). This is the only species to really detain birders in this area, so it is not essential to stay here.

*Location*

From Guánica, take highway PR116 west to highway PR304, where you should turn south, which goes direct to the parador. From highway PR2 take the turn onto highway PR320 south through San Germán. From here, to get to Villa Parguera take the PR101 and then the PR116 at Lajas, before taking the same turning off the PR116 to La Parguera.

***Accommodation*** The two paradors in La Parguera are ideal places to stay *and* see the blackbird. The Villa Parguera is the slightly more pricey of the two, but has a nice restaurant and gardens, which are frequented by the blackbird. The birds roost in the nearby mangrove. The hotel is on the PR304 (telephone: 787-899-7777; e-mail: pvparguera@prtc.net; website: www.villaparguera.net)). Nearby is the Posada Porlamar, which is a bit more reasonable but not as nice (telephone: 787-899-4015) and seemed to be shut as of March 2009. Should you not elect to stay overnight in this area, it is possible to enter either of the hotels to see the blackbird. Just make sure you buy a drink or two from the bar. The Villa Del Mar (telephone: 787-899-4265) and Andin's Chalet (telephone: 787-899-0000) are also in this immediate area. Both charge approximately US$100 per night for a double room. If you require budget accommodation, go inland to Lajas. Among the smaller guesthouses is Villa Andujar Guest House, HC 01-Box 6819 (telephones: 787-899-3475/899-8346. There are a few others in the same area in a similar price range. Options in San Germán include the Parador Oasis, 72 Luna Street (telephone: 787-692-1175), and there is also the Villa del Rey on highway PR360.

***Strategy*** Although the blackbird can be found in small numbers in the gardens of the hotels mentioned above, especially early morning or late afternoon, as of spring 2009 the best bet for this species was to continue on the beach road past the Villa Parguera, with the mangroves on your left, until you see a general store on your right, with a few trees around it and a parking lot behind. The owners put out food daily (usually in the late afternoon), attracting a range of species including groups of up to 20+ Yellow-shouldered Blackbirds. There is also a birdbath, meaning that there nearly always some birds around the area. If you continue to the end of this road, which terminates in a construction site and a small sewage farm, there are chances for a few wetland species, even including White-cheeked Pintail here.

***Birds*** In addition to Yellow-shouldered Blackbird, the following species are possible in the immediate vicinity: Yellow-crowned Night Heron, Eurasian Collared Dove (introduced), Black Swift, Antillean Mango, Puerto Rican Emerald, Puerto Rican Woodpecker, Grey Kingbird, Caribbean Martin, Yellow Warbler, Puerto Rican Spindalis, Pin-tailed Whydah (introduced), Black-faced Grassquit and Greater Antillean Grackle. Multiple introduced parrot species, including Canary-winged Parakeet and Hispaniolan and Orange-winged Parrots, can be seen in the parking area of the Inter American University, just west of San Germán. Mid-afternoon is apparently best. The campus is reached by exiting highway PR2 at San German, proceeding south to highway PR102, turning right and continuing to the campus, in front of which there are several fast-food restaurants lining the street. The gate is on the right. Drive past the athletic fields on the right and at the T-junction turn left. The road winds down to a track with covered bleachers.

## Cabo Rojo and Cartagena Lagoon

*(see Site 11 on main Puerto Rico map)*

If you have some extra time and haven't seen the introduced Venezuelan Troupial, you may want to continue on from Lajas to the Cabo Rojo National Wildlife Refuge in the extreme south-west of Puerto Rico. Yellow-shouldered Blackbird is resident here as well (about 80% of the total population occurs between La Parguera and Cabo Rojo, and is subject to a government recovery programme under the aegis of the Department of Natural and Environmental Resources). The salt flats along the way to the Cabo Rojo lighthouse are great for shorebirds, especially during passage periods, while herons and warblers are common in the coastal mangroves, and Brown Booby can be seen on the rocks below the lighthouse. The lake (also a designated National Wildlife Refuge) is also a productive area, which regularly turns up interesting and unusual birds for the island.

***Location*** Follow the PR101 west from Lajas or the PR100 south from Mayagüez. At Las Arenas look for a sign marked 'El Combate' and turn south here onto the PR301, following it to the salt flats on the left (15.5km). The area of mangrove lies beyond the flats, and the road terminates at the lighthouse. Cartagena Lagoon can also be reached from either Lajas or Mayagüez. From the former, take the PR101 as far as Llanos and turn south at km12.2 on the rather rough PR306. To reach the Boquerón State Forest and avian refuge (which closes at 16.00), continue on the PR101 beyond Las Arenas and take the left-hand turn before the town of Boquerón.

***Accommodation*** Most birders are likely to stay at one of the other localities mentioned for this part of the island. There is no particularly handy accommodation in the immediate vicinity of these two birding areas, other than the Parador Bahia Salinas close to Cabo Rojo, where you can also get a meal.

***Strategy*** This area of the island is not especially productive for endemics, but anyone visiting Puerto Rico for more than a couple of days will certainly find it interesting to bird these sites, especially in winter or during passage periods. Many rarities have been reported in this region, including one of the first records of Western Marsh Harrier for the Western Hemisphere, while the only record of Antillean Palm Swift for the island was seen flying around the cliffs at Cabo Rojo. The 428ha Cartagena Lagoon was formerly an excellent freshwater marsh, but in recent years siltation has occurred and large areas have become dominated by cattails. Nonetheless, reasonable birding is possible here, particularly at the western end, and West Indian Whistling Duck is still present in the area (at least 30 birds), even being seen at times of day other than early morning or late afternoon. The US Fish & Wildlife Service has been actively engaged in restoring the area to its former glory, with some success. One point where you can view the lake on the PR306 is at 18°00'674"N 67°06'540"W. It is also possible to park on the PR306 and walk east through a blue gate along a dirt track for about 1km to a wooden boardwalk and recently constructed observation tower, from where you can gain good views of birds arriving to roost for the evening. Yellow-breasted Crake has been seen from here. A number of introduced exotics can be seen in the reeds around the same point.

Return to the PR101 and proceed west and then south along the PR301, which continues (although somewhat pot-holed) all the way to the extreme south-west tip of the island, at Cabo Rojo. Check the dredging pools a few kilometres inland of the lighthouse, as well as the mudflats and mangrove-lined estuary at the point, for shorebirds (including Wilson's and Piping Plovers) and others. There is a wader-viewing tower, although it apparently is closed from 16.00 onwards, and a nearby trail.

For information on Cabo Rojo and Cartagena Lagoon contact the refuge manager at Cabo Rojo NWR, Carretera 301, Km 5.1, PO Box 510, Boquerón PR 00622. The online map (www.fs.fws.gov/caribbean/Caborojo/Maps.htm) may prove useful.

***Birds***

The following species are among the possibilities in this general area: Magnificent Frigatebird, Brown Booby, a wide variety of herons, West Indian Whistling-Duck, White-cheeked Pintail, Masked Duck, with many other ducks possible, especially in winter and early spring, Osprey, Purple Gallinule, Clapper Rail, Sora, Yellow-breasted Crake, American and Caribbean Coots, Wilson's and Piping Plovers (the latter a rare winter visitor), Royal Tern, Antillean Mango, Caribbean Martin, Cave Swallow, Yellow Warbler, Yellow-faced Grassquit, Yellow-crowned Bishop, Nutmeg and Bronze Mannikins, Venezuelan Troupial and Yellow-shouldered Blackbird (which can be watched coming into roost from the Parador Bahia Salinas).

## Río Abajo State Forest

This protected area and Important Bird Area is situated immediately west of highway PR10, and is roughly equidistant between Utuado (see below) and the coastal city of Arecibo, which in turn is about one hour west of the island's capital San Juan. The main reason for visiting is to see the globally threatened Puerto Rican Parrot, although a good range of the island's other endemics can also be found in this area. If you wish to visit the general area of the Puerto Rican Parrot reintroduction programme, which is managed by the scientific department of the Ministry of Natural Resources in cooperation with the US Fish & Wildlife Service, then it is essential that you endeavour to make advance arrangements by telephone. The reintroduction effort has been in progress for about 20 years, and at the time of writing there were

30–40 free-flying birds (many with radio transmitters attached to them) in the vicinity of the aviaries, which contain approximately 120 Puerto Rican Parrots and about 20 Hispaniolan Parrots, which are used as 'foster parents'. The free-flying birds have started to breed themselves, away from but still in close proximity to the aviaries in recent years.

***Location*** Leave the PR10 westwards, following the sign to the state forest. After less than 1km the road forks around some buildings. The right fork is signed to the visitors' centre, while the left-hand turn is signed to the park offices. The former road, after passing the centre, soon loops back to highway PR10, but the left turn passes through rich forest with many birds.

***Accommodation*** There is no on-site accommodation, with the nearest hotels being in Utuado, to the south, and Arecibo, to the north, with the latter boasting the larger range of options. See **Other sites** for one possibility in the first-named town.

***Strategy*** For those intent only on doing some general birding, then the area to concentrate on is the road leading left, past the park administration, which also leads through a number of open areas with houses as well as closed forest. If, however, your main aim is to see the parrot, then it is imperative that you make access arrangements in advance with the relevant authorities. Telephone the programme on any one of the following numbers: 787-815-1509/-1428/-1517. The same left-hand road close to the entrance of the state forest eventually reaches a yellow barrier, which is usually locked. Beyond this, it is only possible to continue by prior arrangement. The road continues through excellent forest for just over 1km, until a second (electronic) gate is reached. Just beyond here are the aviaries and the birds should be audible at many times of day from this gate. If you have permission to bird this area, then it is at present (spring 2009) best to wait around the second gate around an hour before dusk. Parrots frequently overfly the forest here and, with luck, can even be seen feeding in the trees as well. Access beyond the second gate is only rarely granted and even then not to within sight of the aviaries, to which access is strictly restricted to members of the reintroduction programme. However, if you are able to continue beyond the second gate this will significantly increase your chances of getting good views of the free-flying parrots, including in the early hours of the morning as well. However, such permission is only usually given in special instances, so do not expect to be granted more than access to the area before the second gate.

***Birds*** The following are all possible in this area: Broad-winged Hawk, Scaly-naped Pigeon, Zenaida Dove, Key West and Ruddy Quail-Doves, Puerto Rican Parrot, Mangrove and Puerto Rican Lizard Cuckoos, Puerto Rican Screech Owl, Green Mango, Puerto Rican Emerald, Puerto Rican Tody, Puerto Rican Woodpecker, Puerto Rican Pewee (common),

Puerto Rican and Black-whiskered Vireos, Red-legged Thrush, Pearly-eyed Thrasher, Puerto Rican Bullfinch, Black-faced Grassquit, Puerto Rican Tanager and Puerto Rican Spindalis.

**OTHER SITES**

If on holiday with a non-birding spouse and part of your trip is to tour the city of **San Juan** *(See Site 12 on main Puerto Rico map)*, make sure you visit the old fort (Castillo del Morro), where great looks at White-tailed Tropicbirds are possible along the battlements. This being Puerto Rico, a stop at the nearby Isla Grande Airport, could produce a veritable feast of introductions (probably only of real interest to those with a serious desire to boost their Caribbean list), including Canary-winged Parakeet, Brown-throated Parakeet, Orange-cheeked Waxbill, and Nutmeg and Bronze Mannikins. The Hispaniolan Parrot has also been introduced to Puerto Rico, and can be seen at the USCG base in Old San Juan, where they often permit close approach.

The **Parador Guajataca**, at Quebradillas, on the north coast of Puerto Rico west of Arecibo *(See Site 13 on main Puerto Rico map)*, affords a scenic view of the sea. It has a secluded cove and white sand beach. The rooms have balconies, and the restaurant is excellent and affordable. You can sit on your private balcony, sip a drink and watch White-tailed Tropicbirds to your heart's content, with Red-footed Booby another possibility while seawatching from here. We have not investigated the area personally, but the **Guajataca State Forest**, accessed via the PR446 south-west of Quebradillas is reputedly one of the best sites on the main island of Puerto Rico for Key West Quail-Dove.

Another site solely of value to those obsessed by either their Caribbean or Greater Antilles list is **Ponce airport** *(See Site 14 on main Puerto Rico map)*, just west of the town and south of highway PR52, where colonies of introduced Yellow-crowned Bishop and Indian Silverbill occur.

Situated in the mountainous interior of the island *(See Site 15 on main Puerto Rico map)* are two paradors. At **Jayuya**, on the PR144, near the highest point of the island (the 1,338m-high Cerro Punta), is the Parador Hacienda Gripinas (telephone: 787-828-1717); while the Parador Casa Grande is near **Utuado** (telephone: 787-343-2272), which lies further west on the PR123. Both are sited in abandoned coffee plantations, and are quite delightful, with most of the common endemics on tap in either area. At Casa Grande playing a tape will bring Puerto Rican Screech Owls to your balcony, while the endemic hummingbirds are abundant.

At **Mayagüez** *(See Site 16 on main Puerto Rico map)* on the central west coast of the island, Cayenne Tern, which is sometimes split from Sandwich Tern, occasionally occurs in the harbour. Nearby, at **Punta Algarrobo**, a stretch of beach north of the same town, it is possible to park at the end of Calle Los Alamos (18°15′48″N 67°11′09″W) and walk north on the beach to the outflow of a river, where at low tide, a sandbar is exposed which may hold large numbers of roosting gulls, terns and shorebirds. Some observers have reported several island rarities at this site in recent years, so it is worth checking by those with time at their disposal.

**Desecheo Island** (in the Mona Passage between Puerto Rico and Hispaniola) is a National Wildlife Refuge and has been important for nesting seabirds. Unfortunately, both goats and monkeys have been introduced, and the native vegetation was also badly hit by Hurricane Georges in September 1998. Humpback Whale *Megaptera novaeangliae* is seen offshore, especially in the early months of the year, and great snorkeling is to be had in the environs. The boat ride from Rincón, at the westernmost point of Puerto Rico, takes about 45 minutes and two year-round dive shops near the lighthouse operate boat trips to Desecheo as well as dedicated whale-watching trips.

# CAYMAN ISLANDS

The low-lying Caymans are outcrops of the Cayman Mountain Ridge, an underwater mountain range that extends from Cuba to Central America. The islands are limestone with numerous sinkholes, and covered by shallow-rooted forest. Arguably their most famous avian inhabitant, Grand Cayman Thrush, is now extinct, having last been recorded in 1938. A subspecies of the Jamaican Oriole was extirpated even more recently, and has not been seen since 1967. The total land area of the islands is small: Grand Cayman covers 197km$^2$, Cayman Brac 38km$^2$ and Little Cayman 28km$^2$.

The capital, George Town, is on Grand Cayman, close to the harbour and very near the heavily developed Seven Mile Beach. The Caymans represents one of the major financial centres of the Caribbean, reputedly boasting more registered businesses than people (the human population numbers just over 40,000), who enjoy the highest standard of living in the Caribbean. The government's main source of income is indirect taxation. Tourism accounts for over 70% of annual GNP. The Caymans are renowned as one of the best scuba-diving destinations in the world.

Most visitors stay only a short period and arrive by cruise ship. The Owen Roberts International Airport is a few kilometres east of George Town. If you arrive on a cruise ship and have only limited time ashore, you can actually walk to a good birding area in less than an hour; all of the usual international, as well as locally-owned car rental agencies have offices immediately to the side of the airport terminal. To reach Little Cayman or Cayman Brac you will need to fly, on either Cayman Air or Island Air.

There are relatively few sites to bird in the Caymans, and the only (near-) endemic species is the Vitelline Warbler, which is fairly common (elsewhere, it is also found on the extremely little-visited Swan Islands). For US citizens, Rose-throated Parrot and Cuban Bullfinch will also be of great interest, especially as they are represented by different races on the Caymans. West Indian Whistling Ducks are fairly common here, with a total population of over 2,000 individuals. There are regular (daily) flights between Cuba, or Jamaica, and the Caymans, making it possible to combine either of the two larger islands with the Caymans, or all three archipelagos, in one trip. Note that Cayman Islands' customs requirements necessitate you stating the name of a hotel at which you intend to stay upon entering the country, and that this seems to be taken rather seriously.

## Grand Cayman

### George Town Area (4A, 5A)

If you arrive on Grand Cayman and only have very limited time, the best area for Vitelline Warbler is Walker's Road, which intersects with South Sound Road in the extreme south-west of the island. The warbler can be seen anywhere along this road. This will require a walk of 2–3km from the port, or take a taxi. You may also see Rose-throated Parrot, Northern Flicker (the endemic race *gundlachi*) and Cuban Bullfinch in this area as well. The nearby South Sound Swamp is a haunt of the more localised Yucatan Vireo. Anyone needing accommodation in the island's capital is advised to head for Eldemire's Guesthouse (e-mail: tootie@eldemire.com or telephone: 345-949-5387). Be aware that accommodation is generally rather expensive in the Cayman Islands.

***Map of Grand Cayman***

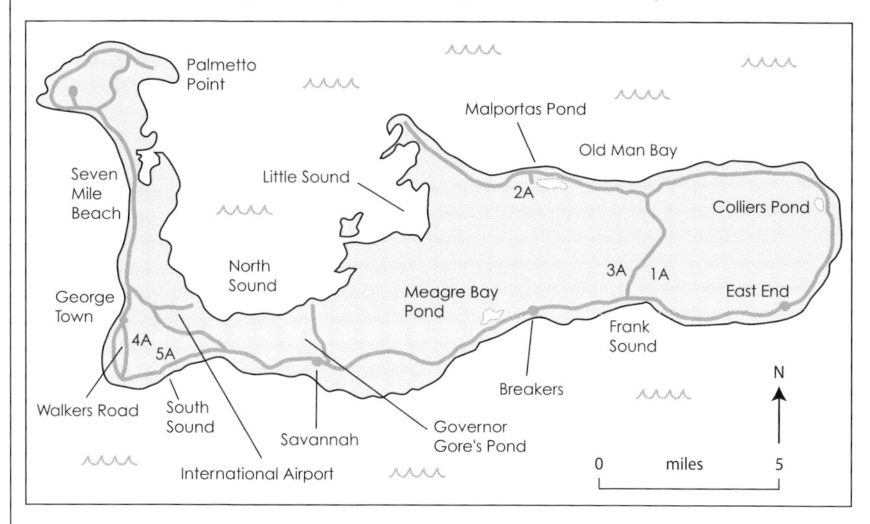


## Botanic Park (1A)

If you are staying on Grand Cayman and have a rental car, arguably the best area is the 50ha Queen Elizabeth II Botanic Park (usually known simply as the Botanic Park). Go east on the A2 past Savannah, Bodden Town and Breakers, then turn left (north) on Frank Sound Road, before turning right (east) into the park. The trip will take about 35 minutes from George Town. There is an entry fee and, as of early 2009, entrance is limited to 09.00 to 17.30. Vitelline Warbler is common here, and West Indian Whistling Duck and Cuban Bullfinch can be seen around the ponds. Other species that can be expected include Rose-throated Parrot, La Sagra's Flycatcher, Loggerhead Kingbird (summer), Caribbean Elaenia, Zenaida Dove, Greater Antillean Grackle (an endemic subspecies), Yucatan and Thick-billed Vireos (both of them represented by endemic races), Western Spindalis (not common), and West Indian Woodpecker (yet another endemic subspecies). Caribbean Dove is a possibility, but the endemic race on the Caymans is rather uncommon. In winter or passage periods migrant warblers may also be present.

## Willie Ebanks' Pig Farm (2A)

After leaving the park, and having returned to Frank Sound Road, turn right and drive about 3km. On reaching the coast keep left, and keep left again when you get to Chisholm's Store. There is a sign for Hutland and 'Mr Willie's Farm' on the left. Drive 1.9km to a dirt road, also on the left. This road reaches the pig farm, which is less than 100m further on, also on the left and very close to where the road reaches a dead-end. Large flocks (numbering hundreds) of West Indian Whistling Ducks come to be fed here, making for a somewhat surreal spectacle.

## Mastic Trail (3A)

This trail is reached by turning to the left off Frank Sound Road 0.6km before you reach the Botanic Park, and immediately after a small fire station. Turn into Mastic Road and continue for another about 1km until you reach the trailhead, which is well signed on the right-hand side. The Mastic Trail is a restored 200-year-old, 4km-long footpath meandering through a reserve of native mangrove swamp, agricultural land and woodland, and goes all the way to the north coast of the island. The highest point on the island, at a mere 18m above sea level, and known locally as 'The Mountain' is traversed en route. The first part is rather wetter and the whole area is a reserve (of 314ha) and an Important Bird Area. Despite some forest clearance at its fringes, mainly for housing developments, this area is a good alternative to the botanic park, boasting a similar range of species, including Mangrove Cuckoo and the endemic Caymans race of the near-ubiquitous Bananaquit, and has the advantage that it does not attract an entrance fee. Vitelline Warbler, Rose-throated Parrot (up to 350 individuals are present in the area), West Indian Woodpecker, La Sagra's Flycatcher and Cuban Bullfinch are all also possible here.

## Governor Gore's Pond

The tiny bird sanctuary at Governor Gore's Pond is well signed off the main Jackson Road just west of the small community of Savannah, west of Bodden Town. Turn north (left if coming from George Town) into Spotts Newlands Road, and shortly afterwards turn right into Pennsylvania Drive and continue to follow this road until you reach a sharp right-hand bend. The entrance to the sanctuary is sited in a layby on the left here. Admission is free. Although none of the main endemic landbirds can be found here, the area is worth a look for anyone with some time to spare, perhaps before a flight. A good range of waterfowl, especially herons (including Least Bittern), passage shorebirds and rails can be found here, while the surrounding vegetation is well worth checking for migrant warblers and others. Any other areas of wetland between here and George Town are worth checking for West Indian Whistling Duck.

# Other Islands

## Cayman Brac

Cayman Brac is a fine area for seeing White-tailed Tropicbird and Brown Booby, both of which can be observed on the 40m-high cliffs at the north-eastern tip of the island. Spot Bay is the best observation point. Vitelline Warbler, of the race *crawfordi*, is common here as well. Red-legged Thrush and, in summer, Antillean Nighthawk can also be found in this area. Caribbean Elaenia and Loggerhead Kingbird are both widespread across this long narrow island. A different race of Rose-throated Parrot (*hesterna*) is present on Cayman Brac and is probably

*May of Cayman Brac*

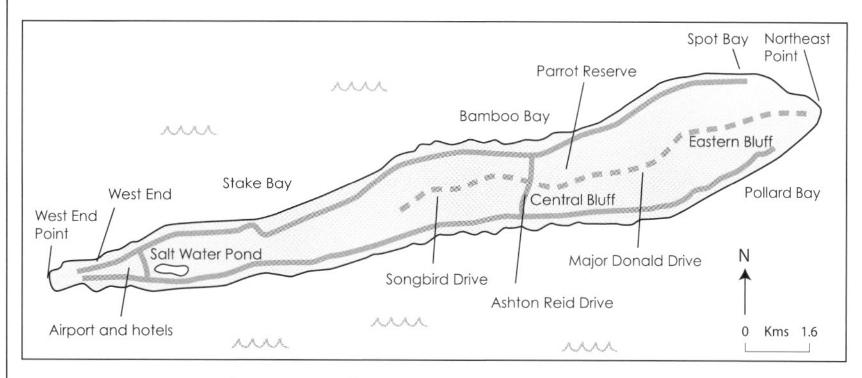


best looked for in the east of the island; it is distinguished by being smaller and darker, with less red on the head than *caymanensis* of Grand Cayman. *The* best area is the Brac Parrot Reserve, a privately owned conservation area along Major Donald Drive, in the vicinity of the Central Bluffs, which also holds the endemic Caymans subspecies of Thick-billed Vireo, *alleni*, and the Near Threatened White-crowned Pigeon. The island is served by a 30-minute flight from Grand Cayman (several per day) and rental cars are available at Gerrard Smith Airport, from where it is about 15km following the north coast of the island to Spot Bay.

## Little Cayman

*Map of Little Cayman*

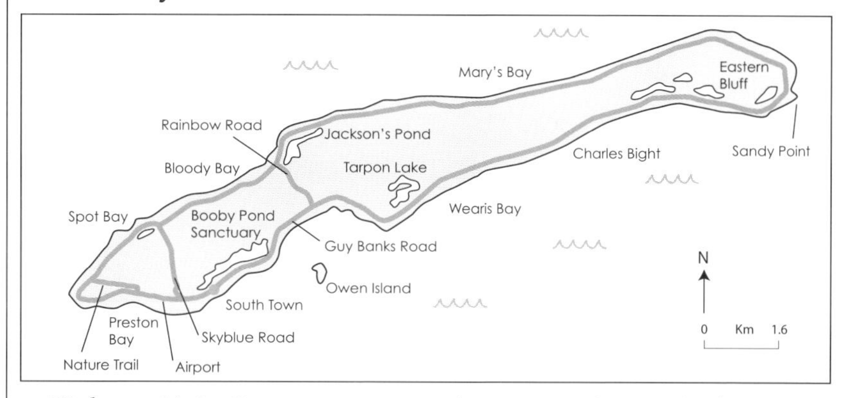


Flights to Little Cayman are currently running about only three times a week (the island is very thinly populated). The main attraction here is the largest Red-footed Booby colony in the Caribbean, sited on the north (landward) side of Booby Pond (43ha), at South Hole Sound, just east of the airport. The colony is partially protected, within the appropriately named Booby Pond Sanctuary, and currently numbers well in excess of 3,000 pairs and perhaps as many as 20,000 individuals (mainly in April to June). This area can also be productive for herons and shorebirds, as well as harbouring West Indian Whistling Duck (best looked for at Jackson's or Grape Tree Ponds), and a breeding colony of Magnificent Frigatebirds. Again, the *crawfordi* race of Vitelline Warbler is easily found here as well, as is White-crowned Pigeon. Another good area for the warbler is Sparrowhawk Hill, an area of relatively pristine and privately owned dry forest in the centre of the island. Access is by tracks on the western and southern sides of the area.

# SELECTED SPECIES ACCOUNTS

**Black-capped Petrel** A globally threatened species (Endangered) that in the non-breeding season wanders north into the waters off the eastern USA. Outside of our region it breeds or has bred on several of the Lesser Antilles (namely Dominica, Guadeloupe and Martinique). Recent work suggests that two different forms occur, 'black-faced' and 'white-faced', which have been speculated to breed at slightly different times, and perhaps on different islands, although the authors of the study were unable to explain the presence of intermediate-plumaged birds. Not easily found in our region, perhaps in part due to the lack of suitable pelagic trips. Enterprising birders might like to investigate the possibility of hiring a suitable boat off southern Hispaniola or eastern Cuba. There are a couple of records of vagrants from the Caymans. ***Cuba*** Speculated to breed in the extreme south-east of the island in part of the Sierra Maestra, but never proven. ***Hispaniola*** The main breeding areas are in the Massif de la Selle and Massif de la Hotte, of Haiti, with smaller numbers in the neighbouring Dominican Republic, on the Lomo del Toro above El Aguacate in the Sierra de Bahoruco.

**West Indian Whistling Duck** A globally threatened species (Vulnerable) that occurs throughout the Greater Antilles. ***Caymans*** Unquestionably the best site in the entire region is Willie Ebanks' pig farm on Grand Cayman, where this duck visits a feeding station in large numbers, but the species can also be found elsewhere in the archipelago, including even in town gardens just south of the capital George Town. ***Cuba*** In Zapata, it is best searched for at Los Canales, but small flocks sometimes perch in the trees alongside the canal banks at Treasure Lake. Also, regularly found in the Sierra de Najasa and Birama Swamp, and, slightly less frequently, on Cayo Coco, particularly around certain of the hotel complexes. ***Hispaniola*** Known in small numbers from mangroves and wooded swamps throughout the island, but best searched for in the Dominican Republic, where the National Botanic Garden in Santo Domingo and Monte Cristí National Park, in the north-west, are two of the best areas. ***Jamaica*** Best searched for at the Black River Morass in the central-south of the island, but numbers are also increasing at the Royal Palm Reserve, in the Negril Great Morass, at the extreme west end of the island. ***Puerto Rico*** Two sites on the main island have proven regular in recent years, the west end of Cartagena Lagoon and the Humacao National Wildlife Refuge. Also worth checking is Caño Tiburones in the north-west of the island. The species was formerly common on Puerto Rico, but has declined dramatically in recent decades, with the current population there estimated at fewer than 100 individuals.

**Cuban Kite** The rarest **Cuban endemic** (Critically Endangered) and now confined to the north-eastern riparian forests, in Moa and Baracoa. Only a few local peasants and a handful of reliable naturalists have seen this raptor during the last decade, although we are aware of additional recent but undocumented reports by visiting birdwatchers. Even during the past three decades only nine ornithologists have seen it: the highly renowned Russian palaeontologist Eugene Kurochkin, the Cubans Nils Navarro, Arturo Kirkconnell, Ernesto Reyes, William Suárez, Alejandro Torres and Orlando Garrido, and the Westerners Guy Kirwan and Mike Flieg. Garrido's sighting, in the late 1960s, occurred at Duaba Arriba, on the bank of the Río Duaba, south-west of Baracoa.

**Gundlach's Hawk** This **Cuban endemic** (Endangered) principally occurs in lower elevation woodland and at the edges of swamps, as well as in montane regions. Found in Casilda (south of Trinidad), Topes de Collantes (north of Trinidad), Cayo Coco, and Gibará (north of Holguín). In Zapata, it is rare, but can be found, with some luck, at several sites, including along the entrance canal to Treasure Lake, between La Boca and Guamá, where it is sometimes seen perched in the introduced *Casuarina* trees that fringe both sides of the watercourse. It has also been seen over the cabins at Guamá and from the tower above the restaurant there, as well as along the road between Playa Larga and Santo Tomás, at Molina, Pálpite, Los Sábalos, Turba, and beside the first part of the road that leads to La Salina. In the area known as Los Canales, it is found, infrequently, at sunset. Wary and elusive, its low population density makes this endangered raptor even more difficult to find.

**Cuban Black Hawk** Also known as the Crab Hawk, this recently split **Cuban endemic** is local but common in coastal habitats: in mangrove, along beaches and in swampy areas where crabs, its main food, are abundant. In Zapata, the best place is the road to La Salina, but scattered individuals are always present elsewhere along the coast (and in times when land crabs are particularly abundant even further inland). Common and most easily found in the Sabana-Camagüey archipelago.

**Ridgway's Hawk** This globally threatened (Critically Endangered) and still-declining raptor is now probably endemic to the **Dominican Republic** (there are no recent reports from Haiti). Sightings were regularly reported in the south-west of the country, in the Sierra de Bahoruco, until recent years (some of which were almost certainly erroneous), but the species now seems to be confined to Los Haïtises National Park, although there are recent reports from the Samaná area. Your best chance of seeing this close relative of the North American Red-shouldered Hawk *Buteo lineatus* is to engage local assistance.

**Yellow-breasted Crake** Widespread but rarely common, virtually throughout the Neotropics, from southern Mexico to southern South America. No records from the Caymans. ***Cuba*** In Zapata, look for it at the entrance to La Cocodrila canal, near the bridge of Los Canales, or at La Turba. ***Hispaniola*** Formerly common over much of the island, but in recent years the best chance of encountering this species has been at Laguna del Rincón, just north-west of Barahona, in the south-west Dominican Republic. ***Jamaica*** Very infrequently recorded, but perhaps not uncommon in the Black River Morass (especially along the Middle Quarters River), in the central-south of the island, and Tryall Golf and Beach Club about 15km west of Montego Bay appears to be a regular site. As elsewhere, seeing one probably depends on luck and watching a suitable area of wetland at dusk or dawn. ***Puerto Rico*** Oberle considers this species to have declined on the island. The only recent records appear to be from the protected areas of Humacao on the east coast and in Boquerón and Laguna Cartagena in the south-west.

**Zapata Rail** This highly localised **Cuban endemic** (Endangered) must be one of the most difficult birds to encounter in the New World. There are scarcely any recent records and the voice remains unknown, as previous statements in the literature (and perpetuated by some local guides) reflect confusion with Spotted Rail. The best chance would be to surprise one in the grassy, swampy marshes north of Santo Tomás or at Peralta. It has also been reported in rather similar habitat at Guamá, in Treasure Lake. Although possible to walk within a few metres of this species, seeing one is almost hopeless due to the height and density of the grass, and the rail's shy nature.

**Spotted Rail** Another widespread rail that is also found on three of the Greater Antilles. Overall, its range spans Mexico to southern South America. ***Cuba*** Quite common in the sawgrass savannas north of Santo Tomás, at Peralta, La Turba and elsewhere in Zapata, but difficult to see. Also found in rice plantations in western Cuba, and recently discovered in the Oriente, near Santiago de Cuba. The best places are the wetlands of Corojal, south-west of Artemisa, and at Santo Tomás. ***Hispaniola*** Extremely rare, being first recorded only in 1978 and to date no regular sites are known for the species. All of the published records are from the Dominican Republic, the most recent in the mangroves of Los Haïtises National Park. ***Jamaica*** Apparently restricted to the Black River Morass, where it is more frequently heard than seen, although it is speculated that the small breeding population might be augmented by winter immigrants.

**Sandhill Crane** This North American species has an isolated population on one of the Greater Antilles. ***Cuba*** The Cuban population of fewer than 1,000 individuals is non-migratory, and birds are smaller than those in North America. Found in small isolated populations, in open savannas or grassy country, where pines are the predominant trees. In Zapata, it could at least formerly be found at Sabanas de San Lázaro, beside the road to Santo Tomás, when the ground was beginning to dry out, and on the road to La Salina, where the species usually foraged in the open grassy area between the forest and mudflats. However, we are unaware of recent records from either area. Also, present in the south-west Isla de la Juventud, parts of the Viñales Valley, north of Itabo (Matanzas province) and several parts of Ciego de Ávila and Camagüey provinces.

**Piping Plover** Restricted as a breeding bird to southern Canada and the

USA, this globally Near Threatened species reaches our region in very small numbers in winter and perhaps also on passage. Not recorded in the Caymans. ***Cuba*** Most regularly found on the cays of the Sabana-Camagüey archipelago, and perhaps best looked for at low tide on Flamingo Beach, on the north side of Cayo Coco, or on Cayo Paredón Grande. ***Hispaniola*** There are a small number of records from just over a handful of sites, most of them in the better watched Dominican Republic, among them Salinas de Baní, but the species cannot be considered regular or common at any of them. ***Jamaica*** A rare winter visitor and no regular sites are known to us, though it has, at least occasionally, been recorded at the Black River Morass in the past. ***Puerto Rico*** Decidedly uncommon on the island and no regular sites are known to us, but it should be searched for on any suitable and relatively undisturbed sandy beaches, including those at Cabo Rojo in the extreme south-west.

**Scaly-naped Pigeon** This pigeon is widespread throughout much of the West Indies, but is unknown from the Caymans and is only a vagrant on Jamaica. ***Cuba*** Though very rare in Zapata, it is common in other regions with royal palm groves and other tall trees. The best places are Soroa and La Güira, but the species certainly occurs on just one cay, Cayo Romano, where it is quite common. ***Hispaniola*** Now largely restricted to moist highland forests in both countries, in Haiti it is only readily found nowadays in La Visite National Park and Macaya Biosphere Reserve. In the Dominican Republic, most visitors are likely to encounter the species in the montane forests of the extreme south-west, in the Sierra de Bahoruco, but it is also regular at Ebano Verde and Los Haïtises National Park. ***Jamaica*** Status poorly known, and the species is apparently restricted to the extreme north-east of the island. ***Puerto Rico*** Reasonably common throughout all of the island's montane forests, being most readily encountered in El Yunque.

**White-crowned Pigeon** Restricted to southern Florida and the West Indies region (including parts of mainland Middle America), this attractive pigeon is still tolerably common throughout the Greater Antilles, although numbers appear to be declining and the species has recently been listed as Near Threatened. ***Caymans*** Generally absent between October and late January on the smaller islands, and much less common on Grand Cayman at this season. At other times the species is reasonably common and widespread in the archipelago. ***Cuba*** The commonest of the larger pigeons, it is very widespread and easily seen at virtually all of the main birding sites on the main island, and is also numerous on several of the satellite islands. ***Hispaniola*** Has declined dramatically, but remains locally common, particularly in mangrove and dry scrub at low elevations. Breeds colonially as elsewhere in the Antilles. ***Jamaica*** Still reasonably common and widespread on the island, this pigeon is readily found in such areas as Windsor Caves, Rocklands Feeding Station, Marshall's Pen, Hardwar Gap, Mockingbird Hill and others. ***Puerto Rico*** Far less common than in the past, some of the only regular modern-day locations are apparently the coastal forests around Dorado, in the north of the island immediately west of San Juan, and between Bayamón and Vega Baja, along highway PR646, also in the north.

**Plain Pigeon** This globally Near Threatened bird is endemic to the Greater Antilles, but is absent from the Caymans. ***Cuba*** The largest and rarest of the island's pigeons. In Zapata, occasionally found beside La Cocodrila canal, at Santo Tomás. Also found near Maneadero, although the best places to see it are the Guanahacabibes Peninsula, in extreme western Cuba; and the Sierra de Najasa, in the area known as La Belén (Camagüey province). It breeds in palms at El Jardín, en route to Vertientes; the environs of the Hotel Tayabito, outside Camagüey city, and it has recently been recorded from the Toa Valley, near Baracoa. ***Hispaniola*** Best searched for in the Dominican Republic, where the dry forests anywhere between Azua and Lago Enriquillo can produce sightings, as can limestone forests in Los Haïtises National Park and Boca de Yuma, in Del Este National Park, and the Alcoa Road in the south-west of the country is another good site. ***Jamaica*** Rare and, until recently, the endemic subspecies *exigua* was even speculated to have become extinct on the island, but birds have recently been seen again in winter in the dry forests around the Portland Ridge, south-east of Lionel Town in the central-south of the island. It reportedly moves to inland forested areas to breed, in April to July. ***Puerto Rico*** Population has increased dramatically in recent years, following near-extirpation, but is still restricted to the eastern end of the Central Highlands, and is best looked for around the town of Comerío.

**Ring-tailed Pigeon** Well named, this **Jamaican endemic** (Vulnerable) is best searched for in mid-elevation forests, particularly anywhere in Cockpit Country, in the north-west, or around the Hardwar Gap, near Kingston, but acquiring good views of a perched individual is not always easy. Also regularly seen around Mockingbird Hill. The species generally descends lower in winter. No other West Indian pigeon has such a prominently banded appearance to the tail.

**Caribbean Dove** Outside our region found only on the small island of San Andrés, in the south-west Caribbean, in south-east Mexico and islands off northern Honduras, with an introduced population on New Providence in the Bahamas. Each island in our region has its own endemic race. ***Caymans*** Uncommon on Grand Cayman, where it is probably best sought in the vicinity of the Botanic Park, and rare on Cayman Brac, where there are only two records, one of them unconfirmed. ***Jamaica*** Reasonably common in lower elevation forests, e.g. of Cockpit Country and regularly seen at Marshall's Pen, outside Mandeville, and at Rocklands Feeding Station.

**Key West Quail-Dove** Found outside our region only in the Bahamas, and is absent from both the Caymans and Jamaica. Generally prefers arid and semi-arid zone forests, but occasionally ascends to wetter and more montane forests. ***Cuba*** Found on the main island, Isla de la

Juventud, especially at Los Indios, and on several cays: Cantiles, Santa María, Coco, Romano, Guajaba and Sabinal. ***Hispaniola*** Best looked for in the south-west of the Dominican Republic, e.g. in the dry forests above Puerto Escondido or south of Barahona. ***Puerto Rico*** Generally rare and infrequently seen, but seems to be reasonably regularly encountered in Susúa State Forest, less regularly in Guánica State Forest, both in the south-west of the island, and also at Laguna Tortuguero, on the north coast west of San Juan, as well at Guajataca State Forest.

**Bridled Quail-Dove** Almost endemic to the Lesser Antilles, this species is rare and not easily found on the easternmost island of our region, and then only by visiting one of its satellites. ***Puerto Rico*** Best searched for on the offshore island of Vieques, although seeing one is not easy, and care must be taken to distinguish this species from Key West Quail-Dove there.

**Grey-headed Quail-Dove** This globally threatened species (Vulnerable) was recently 'split'. The **Cuban endemic** form is less wary than other quail-doves on the same island, and frequently forages along trails and paths. In Zapata, it frequents seasonally wet forests, while Blue-headed Quail-Dove generally prefers drier or limestone areas. Best searched for at Bermejas, Los Sábalos, Molina, La Majagua, Pálpite, and at Santo Tomás en route to La Cocodrila, at the edge of the forest, but also known from several other areas. Zapata is the only region of Cuba where this pigeon is readily found; elsewhere it is the rarest of the quail-doves.

**White-fronted (Hispaniolan) Quail-Dove** Now regarded as a separate species from the Cuban Grey-headed Quail-Dove, this globally threatened bird (Vulnerable) is only certainly known from the Dominican Republic. Note that BirdLife International continues to recognise just one species, so the threat designation refers to *Geotrygon caniceps sensu lato*. ***Hispaniola*** Probably occurs in suitable montane forests of eastern Haiti, but only definitely known from the Dominican Republic, where it is best sought in the forests above El Aguacate in the Sierra de Bahoruco, along the Rabo de Gato trail behind Puerto Escondido, or in the Ebano Verde reserve in the Cordillera Central.

**Crested Quail-Dove** This **Jamaican endemic** (Near Threatened) is regularly observed at Windsor Caves (and elsewhere in Cockpit Country) and Hardwar Gap, as well as at Marshall's Pen, where the owners will usually know the best areas to search at the time. Also found along the Ecclesdown Road in the north-east of the island. This dove is secretive and can be difficult to observe unless you are quiet and paying close attention. No other quail-dove occurs in the highlands of this island, and the population is speculated to be declining.

**Blue-headed Quail-Dove** This **Cuban endemic** pigeon is the sole member of the genus *Starnoenas*, one of four genera restricted to Cuba.

Found from sea level to moderate altitudes, on Pico Turquino, the island's highest mountain, it occurs below 800m in the environs of La Emajagua. This species has suffered greatly from deforestation. However, many remaining woods, especially those on limestone, still harbour this globally threatened species (Endangered). In Zapata, several areas are particularly productive for it: El Cenote, La Majagua, Mera, Molina, Los Sábalos, Bermejas (since 2008, the best area) and Pálpite. At Los Sábalos, early morning and late afternoon, it can be observed walking along the main road. Other good areas to find it are La Güira and the Guanahacabibes Peninsula.

Tips for finding this and the other quail-doves are as follows: a) arrive at your chosen site very early in the morning, b) remain absolutely silent, and c) be careful not to step on dry branches or leaves, to avoid making noise (paths with an abundance of debris or dead leaves are probably going to prove unproductive for this reason). Upon entering the woodland, be alert for their low calls, which once heard make it easier to locate a bird. Without such clues, keep walking the trails, scanning ahead regularly and remaining constantly alert should a bird cross in front of you. All species will traverse roads or paths while foraging. While rather wary and shy, they will sometimes permit very close approach, only taking flight when almost under your feet. Blue-headed Quail-Dove's wings produce a loud and peculiar noise, recalling that of some partridges. No other Cuban doves produce a similar sound. If a quail-dove is heard calling some distance away, it is best to try to reach the area very silently, searching for the bird not on the ground, but perched among the branches. Most call from exposed perches. If the call suddenly ceases, it is almost certainly quite close. You should immediately stop to avoid alarming the bird. Blue-headed and Grey-headed Quail-Doves usually call from some height, while Ruddy and Key West more frequently call from lower perches. Once familiar with the peculiar call of the Blue-headed (a pronounced *uup-up*, resembling that of Common Ground Dove, but with slower and more accentuated syllables), then your search should prove easier. This call is only given during certain months, most frequently immediately prior to the main breeding season (which probably commences in about April). Permanent depressions with stagnant water are also good places to search for all ground-dwelling doves, as the birds visit these to drink during the hottest hours, late in the afternoon, or very early in the morning.

**Hispaniolan Parakeet** Introduced to Guadeloupe, but otherwise a **Hispaniolan endemic**. Like the next species it has declined significantly (currently classed as Vulnerable) and remains only locally common, even having been extirpated from protected areas such as Los Haïtises National Park. In Haiti it is best searched for only at La Visite National Park and Macaya Biosphere Reserve, but is still reasonably numerous in the Sierra de Bahoruco and Sierra de Neiba, in the south-west Dominican Republic. Ironically, however, perhaps the best site is in the capital, Santo Domingo, where the species roosts in large numbers every evening outside the Hotel Embajador.

**Cuban Parakeet** This **Cuban endemic** (Vulnerable) has steadily decreased over the centre and west of the island, especially in Zapata, where it is now rather uncommon and best searched for at Bermejas. Also, found in riparian woodland in eastern Cuba, especially around the Ríos La Melba and Jaguaní, south of Moa. There are breeding colonies at Topes de Collantes, in the Sierra de Trinidad, La Belén, in the Sierra de Najasa, and occasionally elsewhere. Gregarious, it nests in groups, in palm-tree cavities or abandoned woodpecker holes, and frequents cultivated gardens near unmodified areas, such as at Guamá, near Treasure Lake. Several years ago, large flocks occurred in this area, but only scattered birds are seen now. Other places in Zapata suitable for it are Mera, Los Sábalos, La Majagua and Santo Tomás. They are often very silent when foraging, making it possible to be very close to a flock without noticing its presence; easily discovered as soon as they call. Sometimes, falling petals indicate a foraging flock. If a group takes flight, usually two or three birds remain perched, so always carefully check the trees.

**Rose-throated Parrot** This Near Threatened *Amazona* parrot is restricted to two islands in the Bahamas and the westernmost islands of our region. ***Caymans*** Extirpated on Little Cayman but separate races are found on both Grand Cayman and Cayman Brac and remain reasonably numerous, despite illegal hunting. Best searched for on Grand Cayman in the Hutland / Forest Glen / Mastic area, and can be seen anywhere in dry forest on Cayman Brac. ***Cuba*** Common at Zapata, and can be found almost anywhere in this region, as well as at the Guanahacabibes Peninsula, Isla de la Juventud, Topes de Collantes (Sierra de Guamuhaya), the Sierra de Najasa, and several places in the montane east: the Sierra de Moa, Baracoa, Sierra Maestra, and Sierra del Guaso, north of Guantánamo.

**Yellow-billed Parrot** This **Jamaican endemic** is the commoner of the two parrots that are restricted to the island, and appears to become even more widespread in the non-breeding season, when birds move from forested areas into more open wooded regions. Nonetheless, the species is still categorised as Vulnerable by BirdLife International due to illegal hunting (it is perceived as a pest at orchards in some areas), habitat destruction and taking of birds for the pet trade. Undoubtedly best searched for in Cockpit Country, where it is regularly seen from virtually any of the main north to south roads through the area, but it also occurs in the John Crow Mountains, in the east, and there is a flock in Hope Gardens, Kingston (probably not for the 'purists').

**Hispaniolan Parrot** Introduced to Puerto Rico, but is otherwise **endemic to Hispaniola**, where it outnumbers the other endemic psittacine. Nonetheless, the species has declined greatly and many formerly occupied areas are now abandoned. Hispaniolan Parrot is considered globally Vulnerable. Reasonable numbers can be found in the capital of Santo Domingo, but the species is more easily sought in the montane forests of the extreme south-west Dominican Republic,

Ebano Verde reserve, or Del Este or Los Haïtises national parks. In neighbouring Haiti it clings on in small numbers at Macaya Biosphere Reserve and La Visite National Park.

**Puerto Rican Parrot** A **Puerto Rican endemic** and one of the rarest birds in the world (Critically Endangered). It formerly also occurred on the offshore islets of Culebra, Mona and probably Vieques, as well as St. Thomas in the US Virgin Islands. Abundant when Columbus landed in 1493, with a pre-Columbian population of perhaps one million birds, by 1975 the wild population had decreased to just 13 individuals and was confined to the Luqillo region, before slowly growing to a total of 100, of which 53 were in captivity. But, after Hurricane Hugo in autumn 1989, 50% of the wild population was lost, and to prevent extirpation by a single such catastrophe, the captive population, which is slowly increasing, has been split up. Currently, the best area to search for the species is the Río Abajo State Forest in the north-west of the island, but special access arrangements need to be made in advance. The USFWS Puerto Rico Research Station, in Río Piedras, is conducting a study of the parrot. The roads near the La Mina area of the PR191 were formerly good areas to observe parrots moving along one of their main flyways, which passed along the south leg of the El Yunque trail then down the valley into the main gorge upriver of the restaurant. Try El Yunque early in the morning and return in the evening if you have no success. More recently, a vista off highway PR186 on the western side of El Yunque has been more productive. From highway PR3, turn south (right) onto the PR186, pass a small water tank and then a spectacular waterfall on the left, beyond which on the right is the mirador. There are concrete benches and a broad overlook, which encompasses San Juan and its western suburbs. Parrots have been seen flying over the vista in pairs or small groups over a 20-minute period from about 17.30.

**Black-billed Parrot** This species, the smaller of the two **Jamaican endemic** parrots, is largely restricted to Cockpit Country in the north-west of the island, and is considered scarce in eastern Jamaica, though it is regularly found along the Ecclesdown Road. It is best searched for in the vicinity of Windsor Caves or by driving one of the main north to south roads through the Cockpit Country, and by stopping to scan at any convenient overlook. Like the other native *Amazona*, the species is considered Vulnerable by BirdLife International and is considered to be declining due to similar threats as faced by Jamaica's other endemic parrot.

**Great Lizard Cuckoo** Elsewhere this large cuckoo is found only on some of the Bahamas. ***Cuba*** Rather catholic in its choice of habitat, it can be found in most forested areas, including the fringes of sawgrass marshes. Widespread throughout Cuba and Isla de la Juventud, as well as larger cays of the Sabana-Camagüey archipelago, from Cayo Santa María east.

**Puerto Rican Lizard Cuckoo** This cuckoo, which is a **Puerto Rican endemic**, formerly bred on the satellite island of Vieques and possibly on St. Thomas in the US Virgin Islands. It occurs in thickets, forests and shade coffee plantations throughout Puerto Rico, but is most reliably found in Guánica State Forest, at Guajataca, Cambalache or Vega State Forests, or along the entrance road to the El Portal visitor centre in El Yunque.

**Hispaniolan Lizard Cuckoo** This **Hispaniolan endemic** is by far the commoner of the two cuckoos restricted to the island, being generally common in all types of wooded areas, including gardens, to 2,200m. It is most frequent below 1,700m and there is some evidence of local declines, e.g. in Los Haïtises National Park, but the species is usually not difficult to encounter, even occurring in the Santo Domingo Botanic Gardens.

**Jamaican Lizard Cuckoo** Like all lizard cuckoos, this **Jamaican endemic** is more frequently heard than seen, and it is best sought in the vicinity of Hardwar Gap, in Cockpit Country (for instance at Windsor Caves), or at Marshall's Pen, outside Mandeville, as well as in other areas (it is known from suitable habitat virtually throughout the island). Nonetheless, it is probably outnumbered by the even larger Chestnut-bellied Cuckoo, although the lizard cuckoo is to some extent replaced by the *Hyetornis* at higher elevations, especially above 1,200m.

**Chestnut-bellied Cuckoo** One of two species of *Hyetornis* cuckoos in the Greater Antilles, this species is endemic to **Jamaica**, where it is a shy resident of interior and upland forested regions, and its overall range is thus more restricted than the previous species. Like the Hispaniolan Bay-breasted Cuckoo, the species is much more frequently heard than seen, but in comparison to the next species it is rather more widespread and easy to see. There are many regular localities, including Marshall's Pen, Rocklands Feeding Station, Windsor Caves, Ecclesdown Road and Hardwar Gap. It is responsive to playback, but usually not immediately so.

**Bay-breasted Cuckoo** A rare (Endangered) and extremely local **Hispaniolan endemic**, whose status in Haiti is unclear. The only locality where the species has been regularly recorded in recent years that is on the visiting birders' circuit is the Sierra de Bahoruco, in the extreme south-west Dominican Republic, where it is seen with some frequency in the vicinity of 'La Placa' and a couple of kilometres above there, as well as in the region of Rabo de Gato near Puerto Escondido. The species is responsive to playback, but often approaches very quietly and only after some period of time, meaning that some would-be observers 'give up' before the bird arrives!

**Ashy-faced Owl** Although widespread and reasonably common, this **Hispaniolan endemic** is actually one of the most easily missed of the island's specialities. It should also be borne in mind that not every *Tyto* owl is this species, as Barn Owl is also a moderately common breeder on Hispaniola. Consistent areas to find this species are the valleys south of

Barahona, palm-dominated savannas around Monte Plata, north-east of Santo Domingo, and up to 2,000m in the Sierra de Bahoruco, as well as in Los Haïtises National Park.

**Puerto Rican Screech Owl** This **Puerto Rican endemic** is also found on the offshore islands of Culebra and Vieques, as well as formerly on St. Thomas in the US Virgin Islands, where it was last recorded in the 1930s. The subspecies found on those islands other than mainland Puerto Rico appears to be extremely rare and declining due to egg predation by Pearly-eyed Thrasher. The species is found in forested areas, woodlots, treed gardens and edge habitats, and two of the most regular localities are Guánica State Forest and El Yunque.

**Bare-legged Owl** Due to its strictly nocturnal habits, this **Cuban endemic** is very hard to find unless you discover a nesting or roosting hole, usually a natural cavity, old woodpecker nest, or cave. It inhabits wooded areas from sea level to mountains, as well as semi-open country where cabbage palms are common. Look for dead palms with holes; then bang the trunk gently 3–4 times with a stick, while watching the cavity. If the palm is occupied, a bird will usually peer from the hole. Sometimes, it will fly a short distance to the surrounding bushes, or retreat back into the trunk. If one flies out of sight, always search for its mate, which may still be in the cavity. In Zapata, Bare-legged Owl can be observed at Los Lechuzos, Mera, Soplillar, Molina, La Majagua, El Cenote, and especially at Bermejas.

**Cuban Pygmy Owl** The smallest of the island's owls, this **Cuban endemic** is broadly distributed in almost any wooded region. Usually found perched at mid height in trees, they nest in abandoned woodpecker holes or a hollow formed by a broken branch. In places with abundant cabbage palms, it shares these with woodpeckers, trogons and Bare-legged Owl. One or two pairs are always present around the cabins at Playa Larga. Elsewhere in Zapata, it can be found at the mouth of Treasure Lake, Mera, La Majagua, and Bermejas, as well as many other areas. It is active during the day as well as nocturnally.

**Stygian Owl** Widespread, from Mexico to southern South America and also recorded on two islands in our region. During the day, Stygian Owl occupies shady trees, perching at varying heights. At night, however, it is usually perched atop tall trees. Quite rare and usually found in pairs, the members of which may perch a certain distance apart. ***Cuba*** Found on the main island and on Isla de la Juventud wherever there is tall forest. In Zapata, Playa Larga and Playa Girón are the best areas to see it. Once heard, it is easy to find the right tree. Usually, the bird calls from on top of the tallest trees in the garden, from one of several favourite perches, e.g. between the two restaurants or near the discotheque at Playa Girón. Another place in Zapata to see it is at Guamá, in the trees by the Indian village. Away from Zapata, regularly found in the Sierra de Trinidad; at Atkins Gardens, in Soledad, Cienfuegos; and at La Güira. ***Hispaniola*** Unlike on Cuba, this owl is distinctly uncommon here, with no recent records from Haiti. The best areas in the Dominican Republic are the Sierra de Bahoruco, where the species has been heard on the Alcoa Road, and in Los Haïtises National Park.

**Jamaican Owl** This **Jamaican endemic** is accorded its own genus, *Pseudoscops* (one of four endemic genera found on the island). The species is generally quite common and extremely widespread, but local knowledge is extremely useful in helping to locate one. It prefers to trees with many bromeliads and dense foliage, and even when you have tracked one down to a certain tree, the bird can still be difficult to find. Usually the species occurs in pairs. One of the best areas is the garden at Marshall's Pen, where Ann Sutton usually has a nesting pair 'staked out', but Mockingbird Hill is another regular site.

**Antillean Nighthawk** A summer resident in the Greater Antilles, observed from March (exceptionally February, but more usually from April) to October. Quite common, especially on coasts and in pine barrens and open savanna-like habitats. The call, a repeated *que-re-que-te*, immediately identifies it. On migration, especially in September–October and March–April, some of the North American races of Common Nighthawk pass through the region and, unless they are heard calling, the two are virtually inseparable. North American birds emit a nasal *peernt-peernt-peernt*. ***Caymans*** Fairly common to abundant on all three islands, but some summers it may be practically absent from Little Cayman. Regularly seen over Frank Sound Road or downtown George Town, on Grand Cayman. ***Cuba*** Very common on Isla de la Juventud and several cays, including Cayo Largo del Sur. Flocks can be seen at sunset almost throughout this cay. Common at La Güira and Marina Hemingway, close to Havana. In Zapata the species can be observed in coastal areas near Playa Larga, Playa Girón, Pálpite, Soplillar, and Los Canales, about 800m before reaching the first bridge. ***Hispaniola*** Widespread across the island, up to just over 1,100m, even occurring in Santo Domingo. Most regularly hunts over fields, pastures and savannas, such as those above Puerto Escondido, and the area between San Pedro de Macorís and Monte Plata. ***Jamaica*** This island is arguably not the best in the region to search for this species, but regular sites include Marshall's Pen and the Black River Morass, in the summer months (March to October), and the species is widespread across the island in suitable habitat. ***Puerto Rico*** Found over open country, coasts, agricultural fields, pastures and limestone forests, but also over shopping malls! One regular site for this species, which has occasionally even been recorded in winter on the island, is Cabo Rojo.

**Least Poorwill** This globally Near Threatened **Hispaniolan endemic** is perhaps the only extant member of the genus *Siphonorhis* (the other species being found on Cuba and Jamaica). The species' elevational range spans sea level to about 800m, and it prefers dry thorn and cactus scrub (as probably did its congenerics). It is known from only a couple of handfuls of localities, of which just two are in Haiti. For visiting birders, it is probably most easily found in the dry forests between Puerto Escondido and 'La Placa', as well as just south of Barahona. Tape playback is very useful in trying to see this difficult species.

**Cuban Nightjar** A **Cuban endemic** that is common in all forests below 600m above sea level. Its repeated *gua-bai-ro* call is quite distinctive. To find one, enter an area of woodland shortly before sunset and wait for darkness to fall; when a bird starts to sing, approach it. (If you play a tape-recording, the bird may fly in, but it will be hard to discern any field marks on its kestrel-sized shadow). Once located by its call, you can return next day to observe the bird, which usually uses the same roosting perch. Alternatively, drive along suitable dirt roads just after dark, as the species will perch on the roadsides, its bright red eyes visible at some distance in the headlights. In Zapata, the best places to find it are Soplillar, La Majagua, Molina, Bermejas, and Los Sábalos. The most productive roads to check are those to Santo Tomás or Los Canales (before reaching the military fence) or, in eastern Cuba, that between Baitiquirí and Guantánamo, where the bird is very common.

**Hispaniolan Nightjar** This species, which is **endemic to Hispaniola**, is not generally recognised as separate from the previous one. It is fairly common in the west, especially the south-west, of the Dominican Republic, being relatively easily found in the Sierra de Bahoruco above 300m. It does not appear to occur in low-elevation thorn forest. The species is apparently more widespread in Haiti, but probably only easily found in the two principal protected areas in the country.

**Puerto Rican Nightjar** This **Puerto Rican endemic** was thought to be extinct until its rediscovery in 1961, when George Reynard taped an unusual vocalisation in Guánica State Forest. The bird was eventually collected, but it was not until 1973 that information on the species' range and numbers were finally gathered. Puerto Rican Nightjar is presently considered Critically Endangered. The species is currently known from five areas, of which Guánica State Forest, Susúa State Forest and the Guayanilla Hills are the best known. Most observers chose to visit the first-named, where the species is generally very easily found.

**Antillean Palm Swift** This attractive slim-bodied swift is endemic to the three largest of the Greater Antilles, but has wandered to Puerto Rico once. ***Cuba*** Very common on Isla de la Juventud, and in open fields between Soroa and La Güira, in Pinar del Río province; in Havana city, near the beaches; and in Zapata around the cabins at Playa Larga, or at Los Lechuzos, Guamá and Bermejas. ***Hispaniola*** Generally found below 500m throughout the island, but has been reported to over 1,100m in the Sierra de Bahoruco. Distribution largely mirrors that of suitable palms for nesting, but the species was recently discovered breeding in sea caves in the south-west Dominican Republic. ***Jamaica*** As on the other islands, the species' distribution is closely tied to that of palms, and it is principally restricted to lowland regions, although it has apparently declined in the last decade due to hurricane damage.

**Jamaican Mango** A **Jamaican endemic** that is generally common and widespread, especially in lowland and arid coastal regions. Found at many sites, for instance Rocklands Feeding Station, Windsor Caves and Mockingbird Hill. This is the only Jamaican hummingbird that regularly visits cactus flowers to feed.

**Antillean Mango** Basically restricted to two islands in our region; fortunately this multi-island endemic is easily found. ***Hispaniola*** Abundant in dry-forest regions in the lowlands but occurs in most habitats to 2,600m. ***Puerto Rico*** The commonest of the island's five species of hummingbirds in gardens and forested areas along the coast but replaced by the next species at higher elevations.

**Green Mango** This **Puerto Rican endemic** species is the common large hummingbird in the central and west of the island. In the east it is, for instance, somewhat uncommon at El Yunque. It replaces the Antillean Mango at higher elevations, in forests, coffee plantations and edge habitats in the mountains and foothills, and is easily found around Hacienda Juanita, Maricao.

**Green-throated Carib** One of two hummingbirds largely endemic to the Lesser Antilles that just penetrates our region. ***Puerto Rico*** Although this species has been reported from most lowland areas, it is mainly found along the north and east coast as far west as the capital San Juan, and is reasonably easily seen at sites such as Humacao National Wildlife Refuge, as well as on Culebra and Vieques islands. Regularly visits feeders.

**Antillean Crested Hummingbird** Principally restricted to the Lesser Antilles, but just reaches the extreme east of our region. ***Puerto Rico*** Most easily found on the east coast of the main island, for instance at Humacao National Wildlife Refuge and Cabezas de San Juan, as well as on the offshore islands of Culebra and Vieques.

**Cuban Emerald** Despite its name, this species is also found on three of the Bahamas. ***Cuba*** Of Cuba's two resident hummingbirds, this is the larger and more common. It is widespread in woods and gardens, where it regularly visits bottlebrush flowers, and is also common on many of the cays.

**Hispaniolan Emerald** This **Hispaniolan endemic** has become rather rare in Haiti, but remains numerous in the Dominican Republic, where it is generally easily found, although it is commonest in montane habitats, especially above 1,500m.

**Puerto Rican Emerald** This common **Puerto Rican endemic** is most numerous at the western end of the island, but occurs in forests, mangrove and gardens almost throughout the island. However, on the north-east coast of the main island, as well as on the offshore islands of Culebra and Vieques, the species is replaced by the Antillean Crested Hummingbird.

**Black-billed Streamertail** This **Jamaican endemic** is restricted to the eastern third of the island, and is distinguished from its commoner sister-species by the all-black bill, more blue-green plumage and slightly smaller size. Good sites for it include Bath Springs and Mockingbird Hill, at both of which it attends feeders.

**Red-billed Streamertail** The genus *Trochilus* is endemic to Jamaica, and the two species hybridise in the contact zone, which stretches from Port Antonio on the north coast to the Morant River in the south. This **Jamaican endemic** occurs throughout the western two-thirds of the island, where it is generally the commonest hummingbird, and is easily seen at most forested sites, including Rocklands Feeding Station, Marshall's Pen and Hardwar Gap. Small numbers are also found at Mockingbird Hill, permitting observers to study both species together.

**Vervain Hummingbird** This diminutive hummingbird comes close to the Bee Hummingbird in the stakes for the world's smallest bird. It is endemic to two islands in our region. ***Hispaniola*** Represented by the endemic subspecies, *vielloti*, which is reasonably common and widespread, from sea level scrubby woodland, to mixed broadleaf and pine forests at 1,600m. ***Jamaica*** Common and widespread across the island, it is perhaps best seen at one of the hummingbird stations scattered across the island, although it is frequently out-competed at these by one of the larger species.

**Bee Hummingbird** There are at least three places to find this, arguably the most famous **Cuban endemic**: the Guanahacabibes Peninsula, in extreme western Cuba; in the environs of Ojito de Agua and Jaguaní, south of Moa, in Holguín province; and in the Zapata Peninsula. Of these, the best is Zapata, due to its proximity to Havana. Bee Hummingbird is presently treated as Near Threatened by BirdLife International. It is easily mistaken, by the uninitiated, for a bumblebee, owing to its tiny size and the buzzing noise produced by the wings. In areas of tall trees, males almost invariably prefer the highest leafless perches. Territorial, it is very fond of using particular perches for resting or preening. In Zapata, near Molina, a male used the same tree for three consecutive years. Thus, once a bird is located on any of its preferred perches, it will almost certainly be relocated with ease. Despite its diminutive size, it is remarkably loud and is frequently initially located by voice. Both sexes sing. In Zapata, the best localities are La Majagua, La Boca, Bermejas, Santo Tomás, Los Sábalos, Pálpite, and between Molina and La Majagua. At Los Sábalos, it is best to walk the road and search the electric wires.

**Cuban Trogon** This **Cuban endemic** is the country's well-chosen national bird (its colours match those of the national flag) and is still common in most natural forests, and occurs on Isla de la Juventud, Cayo Sabinal and Cayo Guajaba off the north coast. At Playa Larga and elsewhere in Zapata it is found in dry woodland; at La Güira the species is easily located in the pines around the derelict cabins. Prefers shady areas, but is reasonably frequently observed in the open, including areas of sea grape.

**Hispaniolan Trogon** One of only two members of the genus *Priotelus* (the other being the Cuban species), this **Hispaniolan endemic** is most easily sought in the Dominican Republic, having become strangely uncommon in La Visite National Park, although it remains reasonably numerous in Macaya Biosphere Reserve. Good sites for this attractive species include the roads into the Sierra de Bahoruco above 'La Placa' and Las Mercedes, as well as the Ebano Verde reserve.

**Cuban Tody** A usually quite common **Cuban endemic**, and in some places, like Cayo Coco, abundant. Widespread, and known from many locations, it is reasonably common in Zapata, where it best searched for at Los Sábalos, La Majagua, El Roble, and Santo Tomás, as well as Molina, Bermejas and El Cenote.

**Broad-billed Tody** This family, which is endemic to the Greater Antilles, reaches its peak diversity on Hispaniola, which is the only island to harbour more than one species of tody. Of the two **Hispaniolan endemic** todies, the Broad-billed Tody appears to be the most abundant and widespread, being recorded from sea level to at least 1,700m and in most major habitat types.

**Narrow-billed Tody** This **Hispaniolan endemic** is best distinguished from the previous species by voice and bill pattern; it is broadly sympatric with Broad-billed Tody above 1,000m, but species does not appear to occur much below 900m. It is distinctly less easily seen than Broad-billed Tody, but visitors to the montane forests of the south-west Dominican Republic should have no problems finding it, especially if they familiarise themselves with its voice. However, recent genetic evidence suggests that Narrow-billed Tody might actually comprise two species, north and south of the Neiba Valley. If such a split becomes generally recognised, a good place to find the northern 'species' is the Ebano Verde reserve.

**Jamaican Tody** This species, which is **endemic to Jamaica**, is common and easily found on the island. Any area of natural woodland (including mangrove) with a reasonably dense understorey and some earth banks for nesting should harbour the species.

**Puerto Rican Tody** Like all todies, this **Puerto Rican endemic** is more frequently heard than seen, although its *beep-beep* call provides an excellent clue as to the species' presence in an area. It is reasonably ubiquitous in wooded areas on the island, including drier lowland forests, such as Guánica State Forest in the south.

**Antillean Piculet** This **Hispaniolan endemic** is the only piculet in the West Indies and is assigned its own genus. It appears to prefer dry forests and even thorn scrub at middle elevations, but also inhabits coastal mangrove and mixed pine and broadleaf forest as high as 1,770m. In Haiti it is best searched for in the country's two major protected areas, Macaya Biosphere Reserve and La Visite National Park, but is distinctly more widespread and easily sought in the Dominican Republic, for instance along the road between Puerto Escondido and El Aguacate in the extreme south-west.

**Puerto Rican Woodpecker** This **Puerto Rican endemic** is one of three *Melanerpes* woodpeckers that occur in our region. Like the other two species, on Hispaniola and Jamaica, it is common and widespread in wooded habitats, and also occurs on the offshore island of Vieques, where it is less numerous. It formerly occurred on St. Thomas in the US Virgin Islands as well. Its nesting holes may be subsequently used by two other endemics, the Puerto Rican Flycatcher and Yellow-shouldered Blackbird.

**Hispaniolan Woodpecker** As with the next species, this **Hispaniolan endemic** is the only woodpecker to breed on its native island. It is widespread and common, and easily found, even in Santo Domingo.

**Jamaican Woodpecker** The only woodpecker on the island, except for the uncommon Yellow-bellied Sapsucker, which is a winter immigrant, this **Jamaican endemic** is widespread and numerous, and encountered at all elevations and in a wide variety of wooded habitats including gardens.

**West Indian Woodpecker** Outside our region, this species is found only on two of the Bahamas. ***Caymans*** Fairly common on Grand Cayman, it is not present on either of the other islands in the archipelago. Represented by an endemic race. ***Cuba*** Widespread: in Zapata, regularly observed at Playa Larga, Guamá, Playa Girón, Bermejas, Soplillar, Los Lechuzos and many others areas; in Pinar del Río, it occurs at Soroa, La Güira and Guanahacabibes.

**Cuban Green Woodpecker** A common and widespread **Cuban endemic**, this is the smallest and among the most attractive of the island's five resident woodpeckers, as well as being also one of the commonest. Its habitat ranges from dense forest and more open woodland to mangrove. Usually forages in pairs, with the members in close proximity. Its call is a very distinctive, harsh *hurr, hurr* or *tajá, tajá*. In addition, it makes a peculiar clapping sound in flight.

**Fernandina's Flicker** This localised **Cuban endemic** has apparently always been rather rare (the species is currently considered Vulnerable). It prefers open mixed vegetation rather than dense woodland, and is often found in single rows of tall trees. Occasionally social, nesting in close-knit colonies, at other seasons it occurs in pairs and is more difficult to locate. Forages mainly on the ground, and is very fond of ants. At Soroa, often found feeding on the grass as well as on the eaves and roofs of the cabins. Some vocalisations are very similar to those of

Northern Flicker, with which it could be confused in areas such as Bermejas and Soroa, where both occur. The alarm call, however, closely resembles Gundlach's Hawk. Fernandina's Flicker nests in royal palms at Soroa and the Sierra del Rosario; in cabbage palms at Bermejas, Los Galleguitos (south of El Jíbaro, south-east Sancti Spíritus province), as well as at El Dorado, near Isabela de Sagua (Villa Clara province). Also occurs in the Sierra de Najasa, and around Gibará, north of Holguín. Should Fernandina's Flicker and West Indian Woodpecker share a palm, the larger cavity is usually occupied by the former but the commoner species will frequently out-compete and oust the flicker. In Zapata, scattered pairs can be observed in several localities: Santo Tomás, near Pálpite, Mera, and La Turba. The best area, however, is still Bermejas and suitable areas beside the road just west of there, towards Cienfuegos.

**Ivory-billed Woodpecker** This remarkable bird, one of the most iconic in the world, is (or was) represented in our region by a separate subspecies. Recent genetic work has revealed reasonably pronounced differences between Cuban *bairdii* and the endemic continental race. ***Cuba*** This large, crow-sized bird is on the verge of extinction (having already probably been extirpated in the USA) and in the last decades has been seen only around Cupeyal, Ojito de Agua, in northern Guantánamo province. The last official reports were in 1986/87.

**Jamaican Elaenia** This attractive little flycatcher is **endemic to Jamaica**, where it principally occurs in wet forests at mid elevations, as well as shade-coffee plantations and dry-land forests. Some of the best sites for the species include Marshall's Pen, Windsor Caves, Mockingbird Hill and the Ecclesdown Road, but the species is a widespread, if unobtrusive, resident.

**Caribbean Elaenia** This species has a curiously disjunct range on small islands in the western Caribbean and from Puerto Rico through the Lesser Antilles to Trinidad and the Netherlands Antilles. ***Caymans*** It is a common or even abundant resident on all three islands of the archipelago, where it is found in both mangrove and dry forest. Good sites for the species on Grand Cayman include South Sound Road, the Mastic Trail and Botanic Garden. ***Puerto Rico*** Despite having only been discovered on the island in 1963, it is widespread but never common in dry scrubby forests, including Guánica State Forest.

**Greater Antillean Elaenia** This species is endemic to two islands in our region, where it is common in montane regions. Vagrants have been recorded on Mona Island, Puerto Rico, and the Caicos Islands. ***Hispaniola*** The endemic subspecies *cherriei* occurs at elevations between 500m and 2,500m, and is speculated to have declined during the 20th century. Nonetheless, it is still easily found in forested areas of the western Dominican Republic, such as the Sierra de Bahoruco and Ebano Verde. ***Jamaica*** The nominate race is endemic to the island, but is much less regularly encountered than the Hispaniolan form, being only regular at Hardwar Gap on the standard birding circuit of the island. However, the species is considerably more widespread in winter.

**Crescent-eyed (Cuban) Pewee** Widespread in Cuba and the Bahamas, with local races on some of the Cuban cays. ***Cuba*** Crescent-eyed Pewee occurs in different habitats, like mangrove and woodlands. In Zapata, it is very common virtually throughout all wooded areas.

**Hispaniolan Pewee** Two subspecies are recognised for this **Hispaniolan endemic**, the nominate found throughout the main island and the race *tacitus* restricted to the Île de la Gonâve, off Haiti. This species is easily found in all forest types across the island, up to 2,000m, making it one of the more frequently encountered endemics. This species was until recently considered conspecific with Crescent-eyed and Jamaican Pewees. Vagrants have been recorded on the Bahamas and on Mona Island (Puerto Rico).

**Jamaican Pewee** Known locally as the 'Little Tom Fool' or 'Stupid Jimmy', in reference presumably to its rather trusting behaviour, this **Jamaican endemic** is locally common and widespread at middle to higher elevations across the island. Known from most suitable forested sites, it is perhaps even more widespread during the non-breeding season when at least some birds descend to lower elevations, down to sea level.

**Puerto Rican Pewee** This **Puerto Rican endemi**c is sometimes still regarded as a subspecies of the Lesser Antillean Pewee *Contopus latirostris*, which is otherwise found on Guadeloupe, Dominica, Martinique and St. Lucia. The form on the latter island, the nominate, which is sometimes mistakenly referred to as *C. oberi*, is also sometimes given specific status. On Puerto Rico it is common in wooded areas, including coastal scrub and coffee plantations, across the west and centre of the island. Regular sites include Guánica and Maricao state forests.

**Sad Flycatcher** This **Jamaican endemic** is the smallest of the three *Myiarchus* flycatchers found on the island, where it is widespread and common, and one of the most frequently encountered of the endemics. In addition to its voice, other distinguishing characters are its small size, dark crown and yellow underparts.

**Rufous-tailed Flycatcher** This flycatcher is the largest and most distinctive of the *Myiarchus* found in the West Indies, where it is **endemic to Jamaica**. It is regularly encountered at sites such as Marshall's Pen, Rocklands Feeding Station and many others, and is easily identified by its large size (giving rise to its local name, 'Big Tom Fool') and unmistakable rufous tail and primaries.

**La Sagra's Flycatcher** Outside our region, this flycatcher (which was formerly treated as conspecific with the next species) is reasonably common across the Bahamas, and is a casual visitor to southern Florida. ***Caymans*** On this archipelago, the species is restricted to Grand Cayman, where it is reasonably common. Good sites for it include the Mastic Trail. ***Cuba*** Common on the main island, the Isla de la Juventud and on some larger cays: Cayo Largo del Sur and Cantiles, off the south coast, and Francés, Las Brujas, Santa María, Coco, Guillermo, Paredón Grande, Romano, Guajaba and Sabinal, off the north coast.

**Stolid Flycatcher** This *Myiarchus* flycatcher was formerly treated as the same species as La Sagra's Flycatcher, and is endemic to two islands in our region. On both islands it prefers lowland and dry country, including mangroves. ***Hispaniola*** On this island the species seems to perform seasonal migrations, with birds moving higher, to 1,800m, in the breeding season. It is present virtually throughout the island, where it is the only regular *Myiarchus*, making identification straightforward. Frequently seen at most forested sites. ***Jamaica*** Mainly found in dry-forest areas, such as at Rocklands Feeding Station, Portland Ridge and several localities in the wet limestone forests that characterise the north of Cockpit Country. Stolid is larger than the single-island endemic Sad Flycatcher, with a less contrasting dark crown, less extensive yellow on the underparts and more striking wingbars.

**Puerto Rican Flycatcher** Formerly lumped with La Sagra's and Stolid Flycatchers, this species is now considered **endemic to Puerto Rico**, where it is the only regularly occurring *Myiarchus* flycatcher, making field identification straightforward. Its presence is usually obvious due to its loud and persistently given vocalisations, although it sings less frequently outside the breeding season. It is most regularly found in lower elevation dry-forest habitats.

**Loggerhead Kingbird** This West Indian endemic is restricted to our region and some of the Bahamas. ***Caymans*** The race *caymanensis* is most easily found on Grand Cayman, e.g. in the vicinity of the Botanic Park, but is only sporadically present (or extirpated) on Little Cayman. ***Cuba*** The nominate race (*caudifasciatus*) is the commonest and most widespread kingbird on Cuba, where it is easily seen at a great many sites, and also occurs on Isla de la Juventud and several offshore cays. ***Hispaniola*** The race *gabbii* is a recently recommended split. On this island it is generally uncommon and local, especially at low elevations, but it can be found en route into the Sierra de Bahoruco, above Puerto Escondido. ***Jamaica*** The endemic race *jamaicensis* is common and widespread throughout the island. ***Puerto Rico*** Like *gabbii*, the form *taylori*, which is endemic to Puerto Rico and Vieques, has recently been suggested to warrant species status. Unlike other forms, *taylori* has a dark tail tip, and is most regularly found at Maricao State Forest.

**Giant Kingbird** The largest kingbird and, following its extirpation from the Bahamas and perhaps at least one island off Mexico, is now **endemic to Cuba**. Very rare, it is usually seen perched high on quite tall trees, particularly its favoured *Ceiba*. Deforestation has wrought a decline in

Cuba. The best spots currently include La Güira, in Pinar del Río, the Guanahacabibes Peninsula, and especially the Sierra de Najasa, in Camagüey. More widespread in the Oriente, in Holguín province, as well as Pinares del Mayarí, Cabo Cruz, south of Niquero, in Granma province, around Baracoa, and areas of drier forest in the Sierra Maestra, among others.

**Jamaican Becard** This species, the only becard and member of the recently erected Tityridae family to occur in the Caribbean region, is a fairly widespread and tolerably common **Jamaican endemic**. The species' large nests, typical of all becards, provide a good clue as to its presence. It is easily encountered at Marshall's Pen and Mockingbird Hill, but also at many other sites island-wide away from coasts and their immediate hinterlands.

**Cuban Martin** Summer resident (January–September) and **endemic breeder to Cuba**; its wintering grounds are unknown, but are often presumed to be in South America. Males are inseparable from the North American Purple Martin, as they are both entirely blue: in the hand, the Cuban species has inconspicuous white feathering on the lower belly. Females are also difficult to identify in the field, although they have more greyish or brownish feathers on the breast and flanks. North American birds, both female and young, have short streaks on the breast and body-sides, and appear whitish on the underparts. Confusion only occurs when both species coincide in Cuba, in August–September and March–April. In Zapata, a nesting colony can be seen in dead trees on the mudflats, about 200m off the east side of the road to La Salina. At Los Lechuzos, a few pairs nest in the dead palms in the pond. In Old Havana, martins can be observed around the old cathedral and in any of the old churches, in fact in most parts of the city. Any martin seen between May and July is probably this species.

**Caribbean Martin** This species is a summer visitor to most of the West Indies between January and September, presumably moving to South America for the non-breeding season. It is, however, absent from Cuba and is only a very rare passage migrant on the Caymans (records from all three islands), as well as being only a vagrant to the southern Bahamas. ***Hispaniola*** Relatively common from sea level to 1,500m, and is recorded island-wide. ***Jamaica*** Relatively easily seen as on other islands during the breeding season, but may not become widespread on Jamaica until early March (though present from February, until October). ***Puerto Rico*** Absent from late September until January, but otherwise a common island-wide breeding bird, although rarely recorded in large flocks.

**Golden Swallow** This globally threatened (Vulnerable) species is endemic to our region, where it is now probably confined to the island of Hispaniola, with no sightings from Jamaica since 1989. It was common on the latter island into the 1800s, but has been in steady decline since. ***Hispaniola*** The subspecies *sclateri* is now restricted to three mountain ranges in the Dominican Republic and two in Haiti, where it ranges as high as 2,000m. Although regularly recorded in La Visite National Park in the latter country, it is most easily sought in the western Dominican Republic, at Ebano Verde Scientific Reserve, and in the Sierra de Bahoruco, in the pine forests above El Aguacate and along the Alcoa Road, where it breeds in the surroundings of the old bauxite mine.

**Cuban Palm Crow** This **Cuban endemic** is similar to the Cuban Crow, but is smaller, with a stouter, shorter neck. Apparently more widespread in the 19th century, it is nowadays considered to be Endangered and is found only in La Manaja, between Guane and Puerto Esperanza, in west Pinar del Río province, between La Anita and La Mulata in the north of the same province, and in Camagüey province, at Miguel, near Santa Lucía; near the Hotel Tayabito, and El Jardín, San Pablo, and other forested areas near Vertientes.

**Hispaniolan Palm Crow** Treated as a species **endemic to Hispaniola** in this book and the major West Indies field guide, but this split has to date not been recognised by the AOU Checklist Committee. In contrast BirdLife International does recognise the split and treats the species as Near Threatened. Most commonly found in pine forests above 750m, e.g. along the Alcoa Road, which is a regular site, but also at much lower elevations around Lago Enriquillo, where care needs to be taken to distinguish the species from the White-necked Crow, and even occasionally in scrubby habitats around Monte Cristí in the north-west Dominican Republic. Recorded to 3,000m in the Cordillera Central, and is also found in La Visite National Park, Haiti.

**Cuban Crow** This **Cuban endemic** is slightly larger than Cuban Palm Crow. Both species are all black, and although they are distinguished by the differing lengths of their nostril bristles, the best feature is voice: Cuban Crow has very varied vocalisations, with long caws, cackling, and jabbering somewhat reminiscent of a parrot; Cuban Palm Crow possesses a repertoire limited to a repeated, harsh, nasal *kra-kra-kra*, with a somewhat sheep-like quality. Most crows seen in Cuba will be Cuban Crow, which occurs in forests throughout the mainland, Isla de la Juventud, and several larger cays north of Camagüey: Coco, Romano, Guajaba and Sabinal. In recent years, several have been regular around Playa Larga, Zapata.

**White-necked Crow** This **Hispaniolan endemic** and globally threatened species (Vulnerable) is best separated from the Hispaniolan Palm Crow by its wider vocal repertoire and, with particularly good views, its red irides. It is far more widespread across the island than the palm crow, including the extreme east, where it is well known from Del Este National Park and Los Haitises National Park. It is now outnumbered by the palm crow at higher-elevation and pine forests in the south-west Dominican Republic, but can still be found in several areas of this arid part of the country, for instance around Lago Enriquillo.

**Jamaican Crow** A speciality of the Cockpit Country in the north-west of the island, this **Jamaican endemic** is speculated to be expanding its range into agricultural areas. It is the only crow on Jamaica and is most easily found in park-like habitats in the north of Cockpit Country, but is also regular in wetter limestone-based forests and around Marshall's Pen. Also found in the John Crow Mountains, in the east of the island, where it can be found rather easily along the Ecclesdown Road.

**Zapata Wren** This **Cuban endemic** is best searched for in the early morning, before any wind picks up. The birds usually approach quickly, and silently, in response to playback. Often sings from concealed perches, but sometimes boldly approaches in the open, and frequently retreats from its perch by walking through the grass, or on the ground. Zapata Wren was commoner three or four decades ago; it is currently quite rare (being treated as Endangered by BirdLife International), although it appears to have recently colonised at least one area. At Santo Tomás, scattered pairs occupy a large area, and it is currently probably best searched for at Hato de Jicarita or La Turba.

**Cuban Gnatcatcher** This **Cuban endemic** is restricted to the eastern provinces and is very common on the dry southern coast of Granma, Santiago de Cuba and Guantánamo provinces; it also occurs along the north coast of Holguín, around Gibara; further west in Camagüey, at Playa Santa Lucía, and Nuevitas, where it is common; east to the Sierra de Najasa; in the savannas of Bayamo; and also on the cays of the Sabana-Camagüey archipelago (Coco, Paredón Grande, Romano, Guajaba, and Sabinal). Rather uncommon at Paso de Lesca and Los Paredones, in the Sierra de Cubitas, there are also two isolated populations in the centre of the island: the dry area between Cumanayagua and Trinidad, and in Casilda (Los Biasmones). In Casilda, the Hotel Brisas del Ancón lies quite close to Los Biasmones. At Playa Santa Lucía, a number of hotels are located near its habitat. The hotels at Parque Baconao are also convenient. To visit the dry coast east of Guantánamo Bay, it is best to stay at the Hotel Guantánamo.

**Cuban Solitaire** This rather drab **Cuban endemic** is only found in certain montane areas at the easternmost and westernmost extremities of the island. It is much easier to locate in the west, where at least two tourist resorts are sited within its habitat. Due in part to its ventriloquial song, the species is very difficult to locate within the canopy, but it can respond well to playback. The best place is at La Güira, in Pinar del Río province, but it also occurs near Viñales and Rancho San Vicente, as well as more widely in eastern Cuba.

**Rufous-throated Solitaire** This unmistakably beautiful bird also occurs on several islands of the Lesser Antilles. As many as six different subspecies have been recognised, all on different islands. ***Hispaniola*** The endemic subspecies *montanus* occurs virtually island-wide at elevations between 400m and 1,800m. Its distinctive song is a common sound while birdwatching above El Aguacate, in the Sierra de Bahoruco, or at Ebano Verde, but numbers appear to have strongly declined in neighbouring Haiti, for instance in La Visite National Park. ***Jamaica*** Reasonably common breeder in montane forests above 1,000m of the island's interior, for instance at Hardwar Gap and Marshall's Pen, which descends in winter (between December and March) to mid-elevation forests (never to low-lying coastal areas). The endemic race *solitarius* has recently been found to be highly divergent from other members of the species, suggesting that it might be elevated to species status; it is slightly longer tailed and has a brighter throat than *montanus*.

**Bicknell's Thrush** This globally threatened species (Vulnerable), which was only quite recently 'split' from the more widespread Grey-cheeked Thrush, breeds in north-east North America and moves to the Greater Antilles in winter. It is only a vagrant to Puerto Rico, and the species has never been recorded on the Caymans. ***Cuba*** Small numbers have recently been discovered wintering in cloud forest on the highest peaks of the Sierra Maestra in the far south-east of the country. ***Hispaniola*** The island represents the species' major wintering ground. Although recorded in broadleaf forests from sea level, most winter in wet primary forests between 1,000m and 2,200m, especially in the Sierra de Bahoruco, Sierra de Nieba and Cordillera Central, but also elsewhere, including Del Este and Los Haitises National Parks, as well as the main montane regions within Haiti. However, despite the species' relative abundance, which has been revealed by mist-netting studies, Bicknell's Thrush is far from easy to see, although it may prove responsive to playback of the bird's winter vocalisation. ***Jamaica*** Rare and only very infrequently recorded, perhaps only on passage and to date exclusively in the Blue Mountains, in old-growth forests above 1,000m, between September and November, and in March to May.

**White-eyed Thrush** This attractive species is a **Jamaican endemic**, where it is most easily found at Marshall's Pen, Hardwar Gap, and Ecclesdown Road, in the extreme east of the island. It is largely confined to densely vegetated montane regions, but tiny numbers apparently range to lower elevations during the winter.

**La Selle Thrush** One of the most beautiful **Hispaniolan endemics**, this globally threatened (Endangered) thrush superficially recalls an American Robin and is a secretive denizen of montane forests with a dense understorey in both Haiti and the Dominican Republic, at 1,400m to 2,100m. Two subspecies are recognised, but the plumage differences between them are not substantial. The species mainly sings solely at dawn and dusk, when it can choose songposts at the top of small trees, and seeing one at other times of day is largely a matter of pure good fortune. The species is best sought in the forests of the Sierra de Bahoruco, e.g. at Pueblo Viejo, high up on the Alcoa Road, or above El Aguacate, where the species appears to be have become easier to see in recent years, ironically perhaps due to the ongoing deforestation of this area.

**White-chinned Thrush** The second of two thrushes **endemic to Jamaica**, where it is by far the commoner and more widespread *Turdus*. Some appear to move to lower elevations in the non-breeding season, and small numbers are present in the lowlands year-round in the north and south-west of the island, but it is mainly found in any wooded or cultivated area at mid or higher elevations. White-chinned Thrush is perhaps commonest in the Blue Mountains.

**Red-legged Thrush** Endemic to the West Indies; two groups of subspecies are recognised, with the nominate *plumbeus* group in the northern Bahamas, Cuba and the Caymans, and the *ardosiaceus* group on Hispaniola, Puerto Rico and Dominica (on which island the species was perhaps introduced from Puerto Rico). Absent from Jamaica. ***Caymans*** The race *coryi* is endemic to the islands, but only occurs naturally on Little Cayman and Cayman Brac, though a small introduced population survived on Grand Cayman until the mid 1960s. It is most easily found on Cayman Brac, where the species is common on the north coast for instance. ***Cuba*** The western and central race (*rubripes*) has a rufous lower belly, whereas the eastern race (*schistaceus*) is grey, with some white. The only resident Cuban thrush, it is widespread throughout mainland Cuba and occurs on most of the larger cays in the Sabana-Camagüey archipelago. Prefers open woodland and areas with litter, including city gardens. The eastern race occurs at all altitudes in the Sierra Maestra. ***Hispaniola*** The race *ardosiaceus* occurs across Hispaniola and several of its satellite islands (as well as on Puerto Rico), in virtually all habitats and to at least 2,400m. ***Puerto Rico*** A common island-wide resident in wooded regions.

**Bahama Mockingbird** Absent from the Caymans, Hispaniola and Puerto Rico, this mockingbird is now a rare and range-restricted bird on the other two island groups in our region, as well as on many of the Bahamas. ***Cuba*** Rare and restricted to certain northern cays of the Sabana-Camagüey archipelago: Tío Pepe, Lanzanillo, Guillermo (which is perhaps the best area to find it among regularly visited sites on the birding circuit), Coco and Paredón Grande (the second-best locality), but is quite common on Cayo Cruz. ***Jamaica*** The endemic race, *hillii*, is rare in dry scrub in the extreme south of the island, principally in the Hellshire Hills, where the population is speculated to be declining due to recent hurricane damage of its habitat.

**Pearly-eyed Thrasher** This species occurs mainly in the Lesser Antilles and the southern and central Bahamas, where it is speculated to be spreading north. It has been recorded as a vagrant on Jamaica, and appears to have recently spread to the mainland of Hispaniola. ***Hispaniola*** Known from several of the archipelago's satellite islands, which are only infrequently visited by ornithologists, but in 1999 the species was found in the extreme eastern Dominican Republic, in dry scrub forest around Punta Cana. However, it is not necessarily easy to see there. ***Puerto Rico*** Common on the main island, where it regularly predates White-crowned Pigeon eggs and young, and is even found in city gardens and parks. At the start of the 1900s it was rare on Puerto Rico and despite the massive increase in numbers it can still be somewhat secretive and is most easily located by virtue of its raucous calls.

**Palmchat** This **Hispaniolan endemic** is almost inevitably among the first birds to be seen by any birdwatching visitors to the Dominican Republic, almost as soon as they step off the plane. It is ubiquitous, even in towns and cities. The Palmchat is not only an endemic species to the island, but is also placed in its own genus and family. While some may consider its uniqueness to be somewhat devalued by the species' abundance, the Palmchat makes up for this by its enigmatic taxonomy. Although frequently placed as a close relative of the waxwings, its systematic position has yet to be satisfactorily and resolved.

**Thick-billed Vireo** Principally found on the Bahamas, this vireo is otherwise confined to a few of the smaller islands in the Greater Antilles. ***Caymans*** The endemic race *alleni* is still found in reasonable numbers on both Grand Cayman and Cayman Brac, but has been extirpated on Little Cayman. One of the best areas to find this species is the Botanic Park on Grand Cayman. ***Cuba*** An endemic subspecies (*cubensis*) occurs principally on the cays, being reported from Cayo Cinco Leguas, the Sierra Morena (Matanzas province), on Cayo Paredón Grande, where it is most easily found, several times from Cayo Coco, and a few times on Cayo Guillermo, although it is unclear whether resident populations are established on the two latter islands. ***Hispaniola*** The subspecies *tortugae* is a common resident on the Haitian island of Île de la Tortue, but the species has never been recorded on the mainland.

**Jamaican Vireo** This species is one of two vireos **endemic to Jamaica**, and is by far the more readily encountered. Known from the vast majority of regularly visited birding sites on the island, from lowland scrub to montane forest.

**Cuban Vireo** The only vireo **endemic to Cuba**, and more often heard than seen, with the song being easily mistaken by the uninitiated for that of Black-whiskered Vireo (a summer visitor in March–September). Cuban Vireo prefers bushy vegetation, tangles and vines, usually low down. Often occurs in close proximity to either Yellow-headed or Oriente Warblers, and it often forages in mixed flocks with wintering warblers. At Zapata and elsewhere, it is ubiquitous.

**Puerto Rican Vireo** This vireo, which is **endemic to Puerto Rico**, is largely restricted to thorn and scrub forests in the western two-thirds of the island. Nest predation by cowbirds appears to have had an adverse effect on the species, especially after hurricanes, with the resulting habitat destruction permitting the cowbirds to colonise new areas. Nonetheless, it remains reasonably common and easily found (especially once the song is known) in several areas, including Maricao State Forest, and can also be found in El Yunque.

**Flat-billed Vireo** Restricted to the north coast, extreme east and parts of the south of the island, this somewhat unusual vireo is **endemic to Hispaniola** and the offshore Haitian island of Île de la Gonâve. It occurs to 1,200m and is largely confined to dry scrubby habitats. The species is arguably best searched for above and below Puerto Escondido, in the south-west Dominican Republic, or in Del Este and Los Haïtises National Parks, but it can also be seen around Punta Cana, in the same country.

**Blue Mountain Vireo** This **Jamaican endemic** (Near Threatened) is one of the easier species to miss during a short trip to the island. The population is considered to be in decline as a result of habitat change. Despite its name, the species is not confined to the Blue Mountains, but can also be found in the John Crow Mountains and in Cockpit Country, with regular sites including Hardwar Gap, Windsor Caves and Barbecue Bottom, in Cockpit Country, as well as along the Ecclesdown Road, south-east of Port Antonio.

**Yucatan Vireo** Otherwise found on the Caribbean coast of south-east Mexico, Belize and Honduras, this species also has a toehold in our region. ***Caymans*** Found only on Grand Cayman, where the species is represented by an endemic subspecies (the appropriately named *caymanensis*), and has declined in the west of the island. It is best searched for along the Mastic Trail and at the Botanic Park.

**Adelaide's Warbler** This species was recently 'split' into three species inhabiting different islands, Puerto Rico and Vieques (Adelaide's Warbler), Barbuda (Barbuda Warbler) and St. Lucia (St. Lucia Warbler). The **Puerto Rican endemic** Adelaide's Warbler is most easily found at Cabo Rojo and Guánica State Forest in the south-west of the main island, but also occurs at higher altitudes, including around Maricao.

**Olive-capped Warbler** Outside our region, this warbler is confined to pine forests on Grand Bahama and Abaco. ***Cuba*** One of four breeding warblers on the island, it occurs only in the pine country of Sierra de los Organos; and in Pinares de Mayarí, Cupeyal, Montecristo, and several other locations in Guantánamo and Holguín provinces. The best places to find the species are Pinares de Mayarí and, especially, La Güira, where it is common around the disused cabins. The form in Cuba probably represents an endemic subspecies.

**Vitelline Warbler** This species is practically **endemic to the Caymans**; elsewhere, it is found only on the little-visited Swan Islands (owned by Honduras), which lie about 400km to the south-west. Some authors have proposed that Vitelline Warbler should be considered as conspecific with Prairie Warbler of continental North America, but all commentators have been content to admit the race *crawfordi* on Little Cayman and Cayman Brac (nominate *vitellina* occurs on Grand Cayman). The species breeds from late February and is generally common throughout the archipelago, especially on the two smaller islands, where it is a common constituent of the avifauna of low dry woodland and scrub. For those not planning to visit the 'off' islands, and is regularly seen along the Mastic Trail on Grand Cayman.

**Arrowhead (or Arrow-headed) Warbler** This species, which like the next, bears a superficial resemblance to the more familiar Black-and-white Warbler, is **endemic to Jamaica**. Although only locally common, it is not a problematic species to find, being regularly encountered in humid forest at sites such as Marshall's Pen, Hardwar Gap, Rocklands Feeding Station, the Ecclesdown Road, and at various places in Cockpit Country.

**Elfin Woods Warbler** This **Puerto Rican endemic** and globally threatened (Vulnerable) species was discovered as recently as May 1972 and is best searched for in the environs of Maricao or at higher elevations in El Yunque. In 1975 the population of this species was about 450 pairs, in two major areas; by 1983 these populations had declined to 175 birds. However, the present-day population is estimated at 500–900 pairs and the species is threatened by habitat destruction and, especially, nest parasitism by the Shiny Cowbird.

**Yellow-headed Warbler** One of two warblers belonging to a genus **endemic to Cuba**, this species occurs in several scattered locations in western Cuba. Besides Pinar del Río province, where it is quite widespread, the best places are the forests of Zapata. Also occurs on Isla de la Juventud and Cayo Cantiles. The extreme north-eastern range is La Palma, near Itabo, in Matanzas province, which is also the western extremity of the range of Oriente Warbler; the two are separated by less than 10km, but occupy different habitats with Yellow-headed Warbler occurring inland and Oriente in coastal areas.

**Oriente Warbler** The other member of a genus **endemic to Cuba**, where it is common in the east and on most of the larger cays off northern Camagüey and Ciego de Ávila provinces. Shares the habits and habitats of Yellow-headed Warbler, and a similar song. All of those areas mentioned for Cuban Gnatcatcher also support the warbler.

**Jamaican Euphonia** This species is **endemic to Jamaica**, where it is arguably one of the most widespread and frequently seen birds restricted to the island. It is found in virtually any area of trees from sea level to high in the mountains.

**Antillean Euphonia** Restricted to just two islands in our region, but considerably more widespread in the Lesser Antilles. ***Hispaniola*** Locally common, but only easily found in the montane forests of the western Dominican Republic, where it occurs to 2,300m but is generally rare below 600m and in drier habitats. ***Puerto Rico*** Far less common than on Hispaniola, here this euphonia is restricted to shade-coffee plantations and other areas of suitable habitat, including lowland scrub; one of the few regular localities for the species is the Hacienda Juanita, near Maricao.

**Green-tailed Ground Tanager** (Green-tailed Ground Warbler) Formerly classified as a warbler, but recently proven (using genetic methods) to be a tanager, this species is **endemic to Hispaniola**, where it is reasonably common in dry scrub including desertic areas, to moist broadleaf forest, from sea level to over 2,900m. Both this and the next species belong to monospecific genera. Mainly found over the central third of the island, it only just extends into Haiti, in the extreme south-east of the country, but is easily found in the neighbouring Dominican Republic, in the Sierra de Bahoruco.

**Hispaniolan Highland Tanager** (White-winged Warbler) Also **endemic to Hispaniola** and also previously considered to be a parulid, this globally threatened (Vulnerable) species is largely restricted to higher elevation forests, from 875m to 2,000m (being rare below 1,300m). Local extirpations appear to have occurred, perhaps especially in Haiti, as a result of habitat clearance, but chances for encountering this lovely bird remain good in the south-west Dominican Republic, e.g. above El Aguacate.

**Western Spindalis** Outside our region this species, which formed part of the former Stripe-headed Tanager, occurs on the Bahamas and on Cozumel, off south-east Mexico. The group name was changed to make the English names less unwieldy (Western Stripe-headed Tanager etc.) and to emphasise that these birds are, like some other North American 'tanagers', not necessarily closely related to most members of the Thraupidae. ***Caymans*** The race *salvini* is endemic to Grand Cayman, where it is reasonably common and widespread, with one record on Little Cayman. ***Cuba*** Not common in Zapata but can be located at El Roble, most frequently at El Cenote, and less often elsewhere; in contrast, it is quite common at La Güira and Soroa. Widespread in Cuba, from sea level to the higher peaks of Sierra Maestra, and also found on Isla de la Juventud and several cays.

**Jamaican Spindalis** Like the next species, this **Jamaican endemic** is largely found in wet forested hills and mountains, being much rarer at lower elevations and close to coasts. Some of the regular localities include Rocklands Feeding Station, Hardwar Gap, Marshall's Pen and Windsor Caves.

**Hispaniolan Spindalis** A forest-based **endemic to Hispaniola**, which is found in suitable areas of cover across the island, principally from 700m to 2,450m, although locally it is found lower. Good areas include La Visite National Park in Haiti and the neighbouring Sierra de Bahoruco of the south-west Dominican Republic.

**Puerto Rican Spindalis** Like the other members of the grouping, this **Puerto Rican endemic** is generally easy to find, in forests and plantations, or even suburban locations, with good sites including El Yunque and Maricao State Forest.

**Black-crowned Palm Tanager** Found virtually throughout **Hispaniola**, except the extreme west of Haiti, where it is replaced by the next species, this endemic is one of the most readily encountered of the single-island endemics. It is practically abundant at all elevations and in most wooded habitats from sea level to 2,500m. This and the next species are the sole representatives of the genus *Phaenicophilus*.

**Grey-crowned Palm Tanager** Occasionally reported from the extreme south-west Dominican Republic, on the road above El Aguacate, but its occurrence has never been verified there, this species is otherwise **endemic to western Haiti**. It is generally common on the Tiburón Peninsula, and on the offshore islands of Île de la Gonâve (race *coryi*) and Île à Vache (race *tetraopes*).

**Western Chat-Tanager** Until recently, the five subspecies and two species of chat-tanagers were all considered members of a single species, and some disagreement persists over the best taxonomic arrangement for these birds. Rather than being divided along an eastern/western split, the two might be better termed southern and northern species. Western Chat-Tanager is **endemic to southern Hispaniola**, where it occurs from the Massif de la Hotte (race *tertius*) and the Massif de la Selle into the Sierra de Bahoruco (race *selleanus*, which is sometimes considered only doubtfully distinct). It occurs at elevations from just over 700m to 2,200m, but one of the best sites is right at the low end of its altitudinal range on the Alcoa Road, in the extreme south-west Dominican Republic. It is also easily found in the region of Zapotén above El Aguacate.

**Eastern Chat-Tanager** The other member of this superspecies **endemic to Hispaniola** occurs in the Sierra de Nieba (race *neibei*) and Cordillera Central (nominate *frugivorus*), where it is best searched for at the Ebano Verde Scientific Reserve, and on the Île de la Gonâve, off western Haiti (race *abbotti*). It also occurs at low elevations on the Samaná Peninsula, in the extreme north-east of the Dominican Republic. The two chat-tanagers belong to a genus, *Calyptophilus*, endemic to Hispaniola. While BirdLife International currently treats them as one species, they still consider the species to be threatened with extinction (ranking it on current evidence as Vulnerable). Should more than one species be generally recognised, then their respective conservation prospects are likely to be considerably more perilous.

**Puerto Rican Tanager** This is another forest-based tanager belonging to a genus endemic to our region. As its name suggests, the species is **endemic to Puerto Rico**, where it is often a mixed-species flock 'leader', and is regularly found near Maricao and in El Yunque.

**Cuban Bullfinch** This species is entirely confined to our region, where it is found on two of the five archipelagos. ***Caymans*** The race *taylori* is endemic to Grand Cayman (a single record from Little Cayman), where it is frequently encountered along the Mastic Trail or in the vicinity of the Botanic Park. Known locally as the 'Black Sparrow'. ***Cuba*** Found throughout the main island, Isla de la Juventud, and on several cays, including Cantiles off the south coast, and Cinco Leguas, Francés, Santa María, Las Brujas, Coco, Guillermo, Paredón Grande, Romano, Guajaba, and Sabinal in the northern Sabana-Camagüey archipelago. In Zapata, it is widespread, being easily found at most sites.

**Cuban Grassquit** Introduced to New Providence in the Bahamas, but otherwise **endemic to Cuba**, this attractive species was formerly common, but is now rather local and even rare in much of western Cuba, due to intensive trapping for the cagebird trade. Only ten years ago, it was easily found in small flocks at several locations, at Viñales, Guanahacabibes, and Zapata, including Soplillar, Los Lechuzos, Molina, and Bermejas. Now, it is only rarely noted in any of these areas, and the only places where it is observed with any frequency in the west and centre of the island are near Viñales, La Güira and Soroa, the southern Sierra de Trinidad, and the Sierra de Najasa. Another regular site (at least until very recently) lies just outside of Havana, the scrub at the eastern end of Embalse La Coronela, on the south side of the autopista, just west of the turn to El Aguacate and Caimito. However, in eastern Cuba it is still reasonably common and widespread, e.g. in the desertic scrub around Baitiquirí. Usually forms flocks, often with the much commoner Yellow-faced Grassquit, but is typically found in pairs when breeding.

**Yellow-shouldered Grassquit** The genus *Loxipasser*, of which this species is the sole representative, is **endemic to Jamaica**, where it is reasonably common and widespread, although quite easily overlooked. It prefers dry limestone-based woodlands, but is even found in gardens, and is reasonably frequent in parts of Cockpit Country and at Marshall's Pen, but another stronghold is along the Ecclesdown Road in the north-east of the island.

**Puerto Rican Bullfinch** Like the next species, this **Puerto Rican endemic** is a relatively frequently seen species of undergrowth and scrubby areas across the island. It is known from virtually all of the major forested sites that are regularly visited by birders.

**Greater Antillean Bullfinch** Outside our region this species occurs on most of the larger islands of the Bahamas. ***Hispaniola*** As many as three subspecies have been described from Hispaniola and its nearshore islands, but only two are generally recognised, one of which is restricted to Haiti's Île de la Tortue off the north coast (*maurella*) with the other (*affinis*) throughout the main island and all other satellites. Greater Antillean Bullfinch is common and widespread in dry-forest habitats across the main island, from sea level to over 2,000m. It is regularly encountered at many of the Dominican Republic's major birding sites. ***Jamaica*** Likewise, the species (here represented by the subspecies *ruficollis*) is a common and widespread resident in most wooded habitats island-wide, being easily seen at many sites, especially in the mountains.

**Orangequit** Emphasising once again the extraordinary biological riches of the West Indies this **Jamaican endemic** represents yet another genus restricted to our region. It is a locally common nectarivore and is regularly found at Hardwar Gap, Marshall's Pen and Rocklands Feeding Station, among other locations. It is most frequent in mid-elevations forests, but is found in wooded habitats at most altitudes.

**Zapata Sparrow** One of the three famous **Cuban endemics** originally discovered in the early part of the 20th century at Santo Tomás, this species is now increasingly and more appropriately known as the Cuban Sparrow. Thirty years after its discovery, it was found in two other, disjunct areas: east of Guantánamo Bay (around Baitiquirí), and, more recently, at Cayo Coco. The nominate subspecies inhabits sawgrass marshes. In Baitiquirí (subspecies *sigmani*), it occupies an entirely different habitat, xerophytic coastal vegetation. On Cayo Coco (*varonai*), it is widespread in all habitats, and this is the best place to find this unobtrusive bird. It usually feeds on the ground. In Zapata, it is often observed among grass tussocks, but on Cayo Coco, it is observed feeding both on the ground and among bromeliads. The song is a long, wheezy trill, but it also gives a short *chip*. BirdLife International currently treats the species as Endangered.

**Red-shouldered Blackbird** Until recently, most ornithologists considered this **Cuban endemic** to be a subspecies of the North American Red-winged Blackbird. However, it is easily distinguished from all North or Central American populations. Although males have only a slightly different voice and similar plumage to other forms, females are entirely black (not brownish) and lack streaked underparts. Females sing much like the males. Small populations occur only in swampy areas near Itabo (La Lagunita, Salina de Bidos), Matanzas province; at Guanahacabibes, in Laguna de Lugones; and in the Ciénaga de Lanier, on Isla de la Juventud. In Zapata, it is reasonably common, inhabiting sawgrass wetlands, including those north of Santo Tomás, Peralta and La Turba. Separated from Tawny-shouldered Blackbird by its larger size, especially males, the female is entirely black (both sexes of Tawny-shouldered Blackbird have, unsurprisingly, a tawny patch at the wing bend, and adult males possess a scarlet patch bordered by buff and tipped white). The songs of the two species are also clearly different. Young, however, have only a small rusty epaulet admixed black, a pattern confusable with Tawny-shouldered Blackbird, but can be distinguished by two structural features, also common to adults, the squarish tail with slightly pointed tips to the feathers, and thicker based and longer bill.

**Tawny-shouldered Blackbird** Endemic to two islands in the Greater Antilles. ***Cuba*** Common and sometimes occurs in flock with grackles, especially in parks and other man-altered habitats. It prefers open areas, forest edges, and places with trees, especially royal palms, often in suburban areas and city gardens. Found throughout, except the Guanahacabibes Peninsula and Isla de la Juventud. In Zapata, it can be observed in most areas, including Guamá, La Boca (at Treasure Lake), and Playa Larga. ***Hispaniola*** Apart from a single unverified report from the extreme north-west Dominican Republic, the species has been found at just three sites in northern Haiti, where it was initially discovered towards the end of the first third of the 20th century.

**Yellow-shouldered Blackbird** This globally threatened (Endangered) species **endemic to Puerto Rico** is nowadays confined to the extreme south-west of the main island and Mona Island (the latter a separate race, *monensis*). Fortunately, the species appears to be bouncing back from the low of just 200–300 birds in the 1970s, with the population currently estimated at 1,250 individuals and the subject of intensive monitoring and a successful recovery plan. Yellow-shouldered Blackbird is found in dry scrubby woodland and mangroves, in which it roosts, but despite its relative rarity the species is actually rather easy to see in the vicinity of the town of La Parguera, or close to the tip of Cabo Rojo.

**Jamaican Blackbird** This **Jamaican endemic**, known locally as the 'Wildpine Sargeant' or 'Black Banana Bird', is frequently one of the most difficult of the island's specialities to find by visiting birders. Unlike most blackbirds it is neither very vocal nor very social, spending most of its time foraging unobtrusively on bromeliad-laden branches. However, it does occasionally respond well to playback, although usually by simply approaching the sound source rather than vocalising in response. It is relatively widespread in montane regions of the island, at 500 to 2,200m, but is best sought in Cockpit Country, at Hardwar Gap north of Kingston, and at several sites, especially Ecclesdown Road, near Mockingbird Hill, on the north-east coast. BirdLife International treats the species as Endangered, and it is one of the island's four endemic genera (*Nesopsar*).

**Eastern Meadowlark** Widespread across eastern North America, Middle America and northern South America, as well as one island in our region. Recent genetic work suggests that the Cuban population might be meritorious of species status. ***Cuba*** A common resident in open areas throughout the main island, Isla de la Juventud and Cayo Romano. Regular localities for the species include the Sierra de Najasa, near Camagüey, and close to Zapata it can be found in suitable areas beside the Autopista.

**Cuban Blackbird** This common and widespread **Cuban endemic** is, however, absent from the cays and Isla de la Juventud. It nests in palms and coconut trees, and adapts well to man-altered environments, where the species is often easier to find.

**Cuban Oriole** This recently split **Cuban endemic** favours bottlebrush flowers, and is often found around the hotels in Zapata. Found at Los Sábalos, Bermejas, Los Lechuzos, La Boca, Guamá and Pálpite. The species is common at Soroa and on Isla de la Juventud, and is found on several cays: Guillermo, Coco, Paredón Grande, Romano, Guajaba, and Sabinal.

**Hispaniolan Oriole** This **Hispaniolan endemic** is found across the entire island from sea level to at least 1,100m, and is closely tied to the occurrence of palms, being rare in pine-dominated forests and desert scrub. It is arguably commonest in the Cordillera Central and in Los Haïtises National Park in the eastern Dominican Republic, but is also found on several of the larger offshore islands, and is regularly found at many of the other major birding sites.

**Puerto Rican Oriole** The final constituent race of the former Greater Antillean Oriole, which in turn was formerly 'lumped' with the Middle American Black-cowled Oriole. This **Puerto Rican endemic** might be confused with the Yellow-shouldered Blackbird, but has more yellow on the wing and has yellow on the tarsus and ventral region, unlike the blackbird. The oriole is threatened by parasitism by cowbirds, but remains reasonably common and widespread in wooded habitats island-wide.

**Jamaican Oriole** Despite its name, this species formerly occurred on the Caymans (where the last record of the endemic race *bairdi* was in 1967) and is still extant on the little-visited Colombian-owned island of San Andrés off the Panama coast. ***Jamaica*** The species, represented by the endemic race *leucopteryx*, is common and found throughout forested areas of the island, including mangroves and gardens, at all elevations.

**Hispaniolan Crossbill** This globally threatened (Endangered) **Hispaniolan endemic** was for long considered a subspecies of the widespread Two-barred (or White-winged) Crossbill. It is restricted to high-elevation pine forests in the two major mountain ranges in Haiti, as well as the Sierra de Bahoruco and Cordillera Central of the Dominican Republic. Most birders will choose to seek the species in the south-west of the latter country, either above El Aguacate, or at the small drinking pool along the Alcoa Road, which is perhaps the most regular site, but even here it is not guaranteed. There is a single record from Jamaica, at Bellevue in the Port Royal Mountains.

**Antillean Siskin** Despite its name, this species is in fact **endemic to Hispaniola**, where it is mainly confined to the south-west and central mountains of the Dominican Republic and the southern ranges of Haiti, at elevations of 500m to 3,000m. It is probably best searched for in the Sierra de Bahoruco, where the same sites that are best for the crossbill are usually even better for this species, but it also occurs in lower areas too, including at least seasonally at field edges around Puerto Escondido.

# FULL SPECIES LIST

The following bird list, which treats all of the separate nations covered by this book, the Cayman Islands, Cuba, Jamaica, Haiti, the Dominican Republic, and Puerto Rico, is based heavily on the sources listed in the Bibliography and any updates in the recent periodical literature that have come to our attention. Particularly for Cuba, we have also relied on a handful of unpublished data, to make the work a little more complete. For Puerto Rico, an island with an unusually large number of escaped species, many of which have established self-propogating populations in the wild, we have attempted to be as comprehensive in our coverage as possible, by listing all of those species known to us that have apparently been released or escaped. For other islands we have been consciously more selective, because these populations tend to be more poorly known and much less obvious to the visitor. Thus, we are aware of several introductions on Cuba, e.g. Montezuma Quail *Cyrtonyx ocellatus*, and instances of escapes, e.g. Yellow-tailed Oriole *Icterus mesomelas*, none of which has become established.

Taxonomy and nomenclature generally follow those of the American Ornithologists' Union (AOU) Check-list Committee, but the species are ordered following the same system as in the Raffaele *et al*. (1998, 2003) West Indies field guide, to facilitate easy cross-referencing. In a guide such as this, it seems to us more important to produce a user-friendly work, rather than strive to follow the latest higher-level systematics. While we have, for instance, reverted to the genera employed by the AOU in their latest supplement for the species of gulls recorded in the Greater Antilles, we have retained their order in the field guide. We have generally followed the AOU on the issues of names and species limits, but have preferred to follow the sensible (to us) guidelines of the International Ornithological Congress over hyhenation of bird names, and we have deliberately elected to deviate from the AOU in, for instance, recognising two species of Palm Crows in the Caribbean, in treating American (Buff-bellied) Pipit as *Anthus rubescens*, rather than *A. spinoletta*, and in considering Cuban Kite *Chondrohierax wilsonii* to be a species, rather than a subspecies of the widespread Hook-billed Kite *C. uncinatus*. We have also followed Olson & Suárez's recent paper that introduces the genus name *Margarobyas*, in place of *Gymnoglaux*, for the Cuban endemic Bare-legged Owl, because the name *Gymnoglaux* proves to lie in synonymy. The AOU has not yet ruled on this change.

As noted below, species marked in bold type are single-island (or country) endemics. Thus, we have denoted all those species restricted naturally to the island of Hispaniola, even though the vast majority of such taxa occur in more than one country. For our purposes here, we ignore instances of vagrancy proven, or claimed, and introductions in considering endemicity. However, we do treat those few species as endemic whose ranges have contracted through either natural or man-made causes to a single island; Giant Kingbird *Tyrannus cubensis* is one such (and perhaps the best) example. We also treat Cuban Martin *Progne cryptoleuca* as an endemic, because its wintering grounds are effectively unknown, and visiting birdwatchers will therefore certainly wish to see it in Cuba, despite that recent genetic evidence questions whether it is best considered a species apart from Caribbean Martin *P. dominicensis*.

We have also attempted to denote subspecific taxa endemic to the Greater Antilles, sometimes to single islands, in other instances to some or all of the region covered here. Our listing does not pretend to be complete in this respect, for the simple fact that quite a number of subspecies are not necessarily generally accepted by ornithologists. Furthermore, it should be noted that quite a number of subspecific taxa have been named from offshore islands, e.g. off Cuba and Hispaniola, very rarely visited by ornithologists. Not only are these taxa unlikely to be seen by birders, but their validity is often, to some extent, questionable. Finally, it should be noted that we have refrained from coding those single-island or single-country endemic species that are polytypic (i.e. represented by more than one subspecies). In some instances, e.g. Cuban Green Woodpecker *Xiphidiopicus percussus*, quite a number of different races have been named (not all of them presently considered valid). Thus several constraints, principally that of space, have prevented us from elucidating the levels of geographic variation found amongst Greater Antillean birds here.

**STATUS KEY:**

RB = Resident (breeds)
SB = Summer visitor (breeds)
WV = Winter visitor
PM = Passage migrant
OS = Oceanic summer visitor (breeds)
OW = Oceanic winter visitor (breeds)
CV = casual visitor
V = vagrant
H = hypothetical (a somewhat selective listing, but generally where the authors of main works have questioned a record, or the record has never been published and no documentation is to our knowledgde available in its support, we have listed the species in this category)
I = introduced
E = extinct or extirpated
FB = Former breeder
? = Status uncertain

Lower-case abbreviations, e.g. wv, pm, etc., indicate that the relevant species is generally uncommon at the relevant season. This is a necessarily subjective categorisation, which is designed to give only a relatively crude assessment of status.

**Notes:**

Taxa marked with asterisks (*) represent forms (subspecies) that are considered different by some ornithologists and are endemic to parts of the Greater Antilles.

Species marked in **bold** are endemic to single islands/countries.

| English name | Scientific name | Cayman Islands | Cuba | Jamaica | Haiti | Dominican Republic | Puerto Rico |
|---|---|---|---|---|---|---|---|
| Common Loon | *Gavia immer* | | V | | | | |
| Least Grebe | *Tachybaptus dominicus* | V | RB | RB | RB | RB | rb |
| Pied-billed Grebe | *Podilymbus podiceps* | RB, WV | RB, wv | RB | RB | RB | RB |
| Herald Petrel | *Pterodroma arminjoniana* | | | | | | H? |
| Black-capped Petrel | *Pterodroma hasitata* | V | ? | V (FB) | OW | OW | cv |
| **Jamaican Petrel** | ***Pterodroma caribbaea*** | | | **E?** | | | |
| Cory's Shearwater | *Calonectris diomedea* | | V | | H | | |
| Great Shearwater | *Puffinus gravis* | V | | | | V | cv |
| Sooty Shearwater | *Puffinus griseus* | | V | V | | | V |
| Manx Shearwater | *Puffinus puffinus* | | | | | V | wv |
| Audubon's Shearwater | *Puffinus lherminieri* | E | sb | V | | CV | OS |
| Little Shearwater | *Puffinus assimilis* | | | | | | H? |
| Wilson's Storm Petrel | *Oceanites oceanicus* | | V | | cv | cv | cv |
| Leach's Storm Petrel | *Oceanodroma leucorhoa* | | V | V | | V | |
| Band-rumped Storm Petrel | *Oceandroma castro* | | V | | | | |
| White-tailed Tropicbird | *Phaethon lepturus* | OS | sb | rb | rb | rb | rb |
| Red-billed Tropicbird | *Phaethon aethereus* | | V | V | | V | rb |
| Masked Booby | *Sula dactylatra* | V | V | rb | | cv | rb |
| Brown Booby | *Sula leucogaster* | RB | rb | rb | rb | RB | RB |
| Red-footed Booby | *Sula sula* | RB | V | rb | rb | rb | rb |
| Northern Gannet | *Morus bassanus* | | V | | | | |
| American White Pelican | *Pelecanus erythrorhynchos* | V | wv | V | | | V |
| Brown Pelican | *Pelecanus occidentalis* | WV | RB, WV | RB | RB | RB | RB |
| Double-crested Cormorant | *Phalacrocorax auritus* | WV | RB, WV | V | | wv | wv |
| Neotropic Cormorant | *Phalacrocorax brasilianus* | | RB | V | V | H | V |
| Anhinga | *Anhinga anhinga* | | RB | V | V | | |
| Magnificent Frigatebird | *Fregata magnificens* | RB | RB | RB | RB | RB | RB |
| American Bittern | *Botaurus lentiginosus* | wv | wv, pm | V | H | wv | wv |
| Least Bittern | *Ixobrychus exilis* | RB, WV | rb, wv | rb | | | rb |
| Great Blue Heron | *Ardea herodias* | WV, PM | RB, WV | WV | RB, WV | RB, WV | rb, WV |
| Grey Heron | *Ardea cinerea* | | | | | H | |
| Great Egret | *Ardea alba* | WV, PM | RB, WV | RB, WV | RB, WV | RB, WV | RB, WV |
| Little Egret | *Egretta garzetta* | | | V | | V | V |
| Snowy Egret | *Egretta thula* | RB, WV | RB, WV | RB, WV | RB, WV | RB, WV | RB, WV |
| Western Reef Heron | *Egretta gularis* | | H | V | | | V |
| Little Blue Heron | *Egretta caerulea* | RB, WV, PM | RB, WV | RB, WV | RB, WV | RB, WV | RB, WV |
| Tricoloured Heron | *Egretta tricolor* | RB, WV, PM | RB, WV | RB, WV | RB, WV | RB, WV | RB, WV |
| Reddish Egret | *Egretta rufescens* | CV | RB, WV | rb | rb | rb | cv |
| Cattle Egret | *Bubulcus ibis* | RB, WV, PM | RB, WV, PM | RB, WV | RB, WV | RB, WV | RB |
| Green Heron | *Butorides virescens* | RB, WV | RB, WV, PM | RB, WV | RB, WV | RB, WV | RB, WV |
| Black-crowned Night Heron | *Nycticorax nycticorax* | WV, PM | RB, WV, PM | rb, wv | rb, wv | rb, wv | rb |
| Yellow-crowned Night Heron | *Nyctanassa violacea* | RB, WV, PM | RB, WV, PM | RB | RB | RB | RB |
| White Ibis | *Eudocimus albus* | cv | RB | rb | RB | RB | V |
| Scarlet Ibis | *Eudocimus ruber* | | V | V | | | |
| Glossy Ibis | *Plegadis falcinellus* | CV | RB, WV, PM | Rb | RB, WV | RB, WV | cv |
| Roseate Spoonbill | *Platalea ajaja* | cv | RB | V | RB | RB | V |
| Wood Stork | *Mycteria americana* | | | V | | FB, V | |
| American Flamingo | *Phoenicopterus roseus* | V | rb | cv | FB, SV, WV | FB, SV, WV | cv |
| Black Vulture | *Coragyps atratus* | | V | V | | | |
| Turkey Vulture | *Cathartes aura* | cv | RB, wv | RB | RB | RB | I |
| Fulvous Whistling Duck | *Dendrocygna bicolor* | pm | RB, PM | V | rb | rb | rb |
| West Indian Whistling Duck | *Dendrocygna arborea* | RB | RB | rb | RB | RB | rb |
| White-faced Whistling Duck | *Dendrocygna viduata* | | V | | | V | |
| Black-bellied Whistling Duck | *Dendrocygna autumnalis* | V | rb | V | V | | V |
| Tundra Swan | *Cygnus columbianus* | | V | | | | V |
| Greater White-fronted Goose | *Anser albifrons* | | V | | | | |
| Orinoco Goose | *Neochen jubata* | | | H | | | |

| English name | Scientific name | Cayman Islands | Cuba | Jamaica | Haiti | Dominican Republic | Puerto Rico |
|---|---|---|---|---|---|---|---|
| Snow Goose | *Chen caerulescens* | | V | | | | V |
| Canada Goose | *Branta canadensis* | H | V | V | | V | V |
| Brent Goose | *Branta bernicla* | | | | | | V |
| Muscovy Duck | *Cairina moschata* | | I | | | | I |
| Wood Duck | *Aix sponsa* | V | rb | V | | V | V |
| Green-winged Teal | *Anas crecca* | wv, pm | wv, pm | wv | V | V | V |
| American Black Duck | *Anas rubripes* | | H | | | | V |
| Mallard | *Anas platyrhynchos* | I, V | wv, pm | V | | V | V |
| White-cheeked Pintail | *Anas bahamensis* | V | rb | V | rb | rb | rb |
| Northern Pintail | *Anas acuta* | wv, pm | WV, PM | V | wv | wv | wv |
| Garganey | *Anas querquedula* | | | | | | H |
| Blue-winged Teal | *Anas discors* | WV, PM | WV, PM | | WV | WV | WV |
| Cinnamon Teal | *Anas cyanoptera* | | V | V | | | V |
| Northern Shoveler | *Anas clypeata* | wv, pm | WV, PM | wv | wv | wv | wv |
| Gadwall | *Anas strepera* | V | V | V | V | H | V |
| Eurasian Wigeon | *Anas penelope* | | | | | V | V |
| American Wigeon | *Anas americana* | WV, PM | WV, PM | wv | WV | WV | wv |
| Canvasback | *Aythya valisineria* | | V | V | | V | V |
| Redhead | *Aythya americana* | | V | V | | V | |
| Ring-necked Duck | *Aythya collaris* | wv, pm | WV, PM | wv | wv | wv | wv |
| Greater Scaup | *Aythya marila* | H | V | V | | | |
| Lesser Scaup | *Aythya affinis* | Wv | wv, pm | wv | wv | wv | wv |
| Bufflehead | *Bucephala albeola* | | V | V | | | V |
| Hooded Merganser | *Lophodytes cucullatus* | V | V | V | | V | V |
| Red-breasted Merganser | *Mergus serrator* | wv | WV | | | V | V |
| Ruddy Duck | *Oxyura jamaicensis* | V | rb, wv, pm | rb, wv | rb, wv | rb, wv | rb, wv |
| Masked Duck | *Nomonyx dominicus* | V (FB) | rb | rb | rb | rb | rb |
| Osprey | *Pandion haliaetus* | wv, pm | RB, WV, pm | WV | WV | rb?, WV | WV |
| **Cuban Kite** | ***Chondrohierax wilsonii*** | | **rb** | | | | |
| Swallow-tailed Kite | *Elanoides forficatus* | pm | pm | V | V | V | V |
| Snail Kite | *Rostrhamus sociabilis* | | RB | | | | |
| Mississippi Kite | *Ictinia mississippiensis* | | V | V | | | |
| Bald Eagle | *Haliaeetus leucocephalus* | | V | | | | H |
| Western Marsh Harrier | *Circus aeruginosus* | | | | | | V |
| Northern Harrier | *Circus cyaneus* | wv, pm | WV, PM | V | wv | wv | wv |
| Sharp-shinned Hawk | *Accipiter striatus* | | rb* | V | rb* | rb* | rb* |
| **Gundlach's Hawk** | ***Accipiter gundlachi*** | | **RB** | | | | |
| Common Black Hawk | *Buteogallus anthracinus* | | | | | | V |
| **Cuban Black Hawk** | ***Buteogallus gundlachii*** | | **RB** | | | | |
| **Ridgway's Hawk** | ***Buteo ridgwayi*** | | | | **E?** | **rb** | **V** |
| Broad-winged Hawk | *Buteo platypterus* | | RB*, wv | V | | V | rb* |
| Swainson's Hawk | *Buteo swainsoni* | | V | | | V | |
| Red-tailed Hawk | *Buteo jamaicensis* | pm | RB | RB | RB, WV | RB, WV | RB |
| Northern Caracara | *Caracara cheriway* | | RB | V | | | |
| American Kestrel | *Falco sparverius* | WV, PM | RB* | RB* | RB* | RB* | RB* |
| Merlin | *Falco columbarius* | WV, PM | WV, PM | WV, PM | WV, PM | WV, PM | wv |
| Peregrine Falcon | *Falco peregrinus* | WV, PM | rb, WV, PM | wv | WV, PM | WV, PM | wv |
| Red Junglefowl | *Gallus gallus* | | | | | I | I |
| Ring-necked Pheasant | *Phasianus colchicus* | | I | I | | I | I |
| Crested Bobwhite | *Colinus cristatus* | | | | | | I |
| Northern Bobwhite | *Colinus virginianus* | | RB* | | I | I | I |
| Helmeted Guineafowl | *Numida meleagris* | | I | | I | I | I |
| Black Rail | *Laterallus jamaicensis* | | rb?, wv | FB | | rb? | FB, wv |
| Clapper Rail | *Rallus longirostris* | | RB | rb | RB | RB | rb |
| King Rail | *Rallus elegans* | | RB*, pm | V | | | |
| Virginia Rail | *Rallus limicola* | | V | | | | V |
| Uniform Crake | *Amaurolimnas concolor* | | | E?* | | | |
| Sora | *Porzana carolina* | WV, pm | WV, PM | WV | WV, PM | WV, PM | wv |

| English name | Scientific name | Cayman Islands | Cuba | Jamaica | Haiti | Dominican Republic | Puerto Rico |
|---|---|---|---|---|---|---|---|
| Yellow-breasted Crake | *Porzana flaviventer* | | rb | rb | rb* | rb* | rb |
| **Zapata Rail** | ***Cyanolimnas cerverai*** | | **rb** | | | | |
| Spotted Rail | *Pardirallus maculatus* | | rb | rb | | rb | |
| Purple Gallinule | *Porphyrula martinica* | rb, pm | RB, WV | rb | RB, WV? | RB, WV? | rb |
| Common Moorhen | *Gallinula chloropus* | RB, PM | RB, WV | RB | RB, WV | RB, WV | RB |
| American Coot | *Fulica americana* | rb, WV | RB, WV | rb, WV | rb, WV | rb, WV | rb, WV |
| Caribbean Coot | *Fulica caribaea* | | ? | | RB | RB | rb |
| Limpkin | *Aramus guarauna* | | RB | Rb | rb | RB | E |
| Sandhill Crane | *Grus canadensis* | H | rb* | | | | |
| Double-striped Thick-knee | *Burhinus bistriatus* | | | | rb?* | rb* | |
| Northern Lapwing | *Vanellus vanellus* | | | | | | V |
| Black-bellied Plover | *Pluvialis squatarola* | WV, PM | WV, PM | WV, PM | WV, PM | WV, PM | WV |
| American Golden Plover | *Pluvialis dominica* | pm | pm | pm | pm | pm | pm |
| Snowy Plover | *Charadrius alexandrinus* | V | rb | V | RB | RB | rb |
| Wilson's Plover | *Charadrius wilsonia* | WV, PM | SB, pm | RB | RB | RB | RB |
| Semipalmated Plover | *Charadrius semipalmatus* | WV, PM | WV, PM | WV, PM | WV, PM | WV, PM | WV |
| Piping Plover | *Charadrius melodus* | | WV, PM | V? (wv) | | wv | wv |
| Killdeer | *Charadrius vociferus* | WV, PM | RB, WV, PM | RB, WV | RB, WV | RB, WV | RB |
| American Oystercatcher | *Haematopus palliatus* | | rb, wv | V | | rb, wv | rb |
| Black-necked Stilt | *Himantopus mexicanus* | RB, WV, PM | RB, WV, PM | RB | RB | RB | RB |
| American Avocet | *Recurvirostra americana* | V | wv | V | | | wv |
| Northern Jacana | *Jacana spinosa* | | RB | RB | RB | RB | V |
| Common Greenshank | *Tringa nebularia* | | | | | | V |
| Greater Yellowlegs | *Tringa melanoleuca* | WV, PM | WV, PM | WV, PM | WV, PM | WV, PM | WV, PM |
| Lesser Yellowlegs | *Tringa flavipes* | WV, PM | WV, PM | WV, PM | WV, PM | WV, PM | WV, PM |
| Solitary Sandpiper | *Tringa solitaria* | WV, PM | WV, PM | wv, pm | WV, PM | WV, PM | Wv, PM |
| Willet | *Tringa semipalmata* | rb, WV, PM | RB, WV, PM | WV, PM | WV, PM, rb | WV, PM, rb | WV, PM, rb |
| Spotted Sandpiper | *Actitis macularius* | WV, PM | WV, PM | WV, PM | WV, PM | WV, PM | WV, PM |
| Upland Sandpiper | *Bartramia longicauda* | pm | pm | V | | | V |
| Eskimo Curlew | *Numenius borealis* | | | | | H | V |
| Whimbrel | *Numenius phaeopus* | wv, pm | wv, pm | wv, pm | wv, pm | wv, pm | pm |
| Long-billed Curlew | *Numenius americanus* | H | V | V | | | V |
| Hudsonian Godwit | *Limosa haemastica* | | V | | V | V | V |
| Marbled Godwit | *Limosa fedoa* | pm | V | V | V | V | V |
| Ruddy Turnstone | *Arenaria interpres* | WV, PM | WV, PM | WV, PM | WV, PM | WV, PM | WV, PM |
| Red Knot | *Calidris canutus* | pm | wv, pm | | wv, pm | wv, pm | pm |
| Sanderling | *Calidris alba* | WV, PM | WV, PM | WV, PM | WV, PM | WV, PM | WV, PM |
| Semipalmated Sandpiper | *Calidris pusilla* | WV, PM | WV, PM | WV, PM | WV, PM | WV, PM | WV, PM |
| Western Sandpiper | *Calidris mauri* | wv, pm | wv, pm | WV, PM | WV, PM | WV, PM | WV, PM |
| Least Sandpiper | *Calidris minutilla* | WV, PM | WV, PM | WV, PM | WV, PM | WV, PM | WV, PM |
| White-rumped Sandpiper | *Calidris fuscicollis* | PM | pm | pm | V | wv, pm | pm |
| Baird's Sandpiper | *Calidris bairdii* | pm | H | | | V | |
| Pectoral Sandpiper | *Calidris melanotos* | PM | pm | pm | wv, pm | wv, pm | pm |
| Dunlin | *Calidris alpina* | V | V | V | | V | V |
| Curlew Sandpiper | *Calidris ferruginea* | | | | | | V |
| Stilt Sandpiper | *Calidris himantopus* | pm | wv, PM | wv, pm | WV, PM | WV, PM | WV, PM |
| Buff-breasted Sandpiper | *Tryngites subruficollis* | V | V | V | V | V | V |
| Ruff | *Philomachus pugnax* | | | V | | | V |
| Short-billed Dowitcher | *Limnodromus griseus* | WV, PM | WV, PM | WV, PM | WV, PM | WV, PM | WV, PM |
| Long-billed Dowitcher | *Limnodromus scolopaceus* | wv, pm | wv, pm | pm | | pm | wv |
| Wilson's Snipe | *Gallinago gallinago* | WV, PM | WV | wv, pm | WV, PM | WV, PM | wv |
| Wilson's Phalarope | *Phalaropus tricolor* | pm | H | V | | wv, pm | V |
| Red-necked Phalarope | *Phalaropus lobatus* | | V | V | | pm | V |
| Red Phalarope | *Phalaropus fulicarius* | | V | | | | |
| Pomarine Jaegar | *Stercorarius pomarinus* | | wv | V | WV, PM | WV, PM | |
| Parasitic Jaeger | *Stercorarius parasiticus* | | wv | V | WV, PM | WV, PM | pm |
| Long-tailed Jaeger | *Stercorarius longicaudus* | pm | V | V | V | V | |
| South Polar Skua | *Stercorarius maccormicki* | | V | | H | H | V |

| English name | Scientific name | Cayman Islands | Cuba | Jamaica | Haiti | Dominican Republic | Puerto Rico |
|---|---|---|---|---|---|---|---|
| Laughing Gull | *Leucophaeus atricilla* | WV, PM | RB, WV, PM | RB, WV, PM | RB, WV, PM | RB, WV, PM | RB, WV, PM |
| Franklin's Gull | *Leucophaeus pipixcan* | | V | | | V | V |
| Little Gull | *Hydrocoloeus minutus* | | | | | | V |
| Black-headed Gull | *Chroicocephalus ridibundus* | | V | | | | V |
| Bonaparte's Gull | *Chroicocephalus philadelphia* | | wv | | | V | V |
| Ring-billed Gull | *Larus delawarensis* | wv, pm | wv, pm | wv, pm | wv, pm | wv, pm | wv, pm |
| American Herring Gull | *Larus smithsonianus* | wv, pm | wv, pm | wv, pm | wv, pm | wv, pm | wv, pm |
| Lesser Black-backed Gull | *Larus fucus* | | wv | | | V | V |
| Great Black-backed Gull | *Larus marinus* | | wv | | | wv, pm | V |
| Sabine's Gull | *Xema sabini* | | V | | | | |
| Black-legged Kittiwake | *Rissa tridactyla* | | V | V | | V | V |
| Gull-billed Tern | *Gelochelidon nilotica* | cv | wv, pm | wv, pm | wv, pm | wv, pm | cv |
| Caspian Tern | *Hydroprogne caspia* | wv, pm | rb?, wv | wv, pm | WV, PM | WV, PM | V |
| Royal Tern | *Thalasseus maximus* | R, WV | RB, WV, PM | WV, PM | WV, PM | WV, PM | wv, pm |
| Sandwich Tern | *Thalasseus sandvicensis* | wv, pm | SB, WV, PM | WV, PM | WV, PM | WV, PM | sb, wv |
| Roseate Tern | *Sterna dougallii* | | sb, pm | cv | wv, pm | wv, pm | sb |
| Common Tern | *Sterna hirundo* | wv, pm | pm | V | WV, PM | WV, PM | wv |
| Arctic Tern | *Sterna paradisaea* | | V | | | | pm |
| Forster's Tern | *Sterna forsteri* | wv, pm | wv | | | V | V |
| Least Tern | *Sternula antillarum* | SB | SB, PM | sb, WV, PM | sb, WV, PM | sb, WV, PM | sb |
| Bridled Tern | *Onychoprion anaethetus* | os | OS | os | os | os | OS |
| Sooty Tern | *Onychoprion fuscata* | V | OS | OS | OS | OS | OS |
| Large-billed Tern | *Phaetusa simplex* | | V | | | | |
| White-winged Tern | *Chlidonias leucopterus* | | | | | | V |
| Black Tern | *Chlidonias niger* | pm | pm | PM | | pm | pm, wv |
| Brown Noddy | *Anous stolidus* | V | OS | OS | OS | OS | OS |
| Black Noddy | *Anous minutus* | | | V | | | sv |
| Black Skimmer | *Rynchops niger* | cv | wv | V | V | V | wv |
| Dovekie (Little Auk) | *Alle alle* | | V | | | | |
| Rock Dove | *Columba livia* | I | I | I | I | I | I |
| Scaly-naped Pigeon | *Patagioenas squamosa* | | RB | V | RB | RB | RB |
| White-crowned Pigeon | *Patagioenas leucocephala* | RB, PM | RB | RB | RB | RB | RB |
| Plain Pigeon | *Patagioenas inornata* | | rb* | rb* | RB* | RB* | rb* |
| **Ring-tailed Pigeon** | ***Patagioenas caribaea*** | | | **RB** | | | |
| Eurasian Collared Dove | *Streptopelia decaocto* | | RB | | | V | I |
| Ringed Turtle Dove | *Streptopelia risoria* | I | | | | | I |
| White-winged Dove | *Zenaida asiatica* | RB, PM | RB | RB | RB | RB | RB |
| Zenaida Dove | *Zenaida aurita* | RB | RB | RB | RB | RB | RB |
| Mourning Dove | *Zenaida macroura* | pm | RB | RB | RB, wv | RB, wv | RB |
| Passenger Pigeon | *Ectopistes migratorius* | | E | | | | |
| Common Ground Dove | *Columbina passerina* | RB* | RB* | RB* | RB* | RB* | RB* |
| Caribbean Dove | *Leptotila jamaicensis* | rb* | | RB* | | | |
| Key West Quail-Dove | *Geotrygon chrysia* | | RB | | RB | RB | rb |
| Bridled Quail-Dove | *Geotrygon mystacea* | | | | | | RB |
| **Grey-headed Quail-Dove** | ***Geotrygon caniceps*** | | **RB** | | | | |
| **White-fronted Quail-Dove** | ***Geotrygon leucometopia*** | | | | **E?** | **rb** | |
| Ruddy Quail-Dove | *Geotrygon montana* | | RB | RB | RB | RB | RB |
| **Crested Quail-Dove** | ***Geotrygon versicolor*** | | | **RB** | | | |
| **Blue-headed Quail Dove** | ***Starnoenas cyanocephala*** | | **RB** | **I (E)** | | | |
| White Cockatoo | *Cacatua alba* | | | | | | I |
| Budgerigar | *Melopsittacus undulatus* | | | | | I? | I |
| Cockatiel | *Nymphicus hollandicus* | | | | | | I |
| Rose-ringed Parakeet | *Psittacula krameri* | I | ? | | | | I |
| Monk Parakeet | *Myiopsitta monachus* | I | | | | | I |
| Senegal Parrot | *Poicephalus senegalus* | | | | | | I |
| Black-hooded Parakeet | *Nandayus nenday* | | | | | | I |
| Blue-crowned Parakeet | *Aratinga acuticaudata* | | | | | | I |
| **Hispaniolan Parakeet** | ***Aratinga chloroptera*** | | | | **RB** | **RB** | **I (E?)** |

| English name | Scientific name | Cayman Islands | Cuba | Jamaica | Haiti | Dominican Republic | Puerto Rico |
|---|---|---|---|---|---|---|---|
| **Cuban Parakeet** | ***Aratinga euops*** | | **RB** | | | | |
| Red-masked Parakeet | *Aratinga erythrogenys* | I | | | | | |
| Olive-throated Parakeet | *Aratinga nana* | | | RB* | | I | I |
| Orange-fronted Parakeet | *Aratinga canicularis* | | | | | | I |
| Brown-throated Parakeet | *Aratinga pertinax* | | | | | I? | I |
| **Cuban Macaw** | ***Ara cubensis*** | | **E** | | | | |
| Red-and-green Macaw | *Ara chloroptera* | | | | | | I |
| Green-rumped Parrotlet | *Forpus passerinus* | | | RB | | | |
| Canary-winged Parakeet | *Brotogeris versicolurus* | | | | | I? | I |
| Rose-throated Parrot | *Amazona leucocephala* | RB* | RB* | | | | |
| **Yellow-billed Parrot** | ***Amazona collaria*** | | | **RB** | | | |
| **Hispaniolan Parrot** | ***Amazona ventralis*** | | | | **RB** | **RB** | **I** |
| **Puerto Rican Parrot** | ***Amazona vittata*** | | | | | | **RB** |
| **Black-billed Parrot** | ***Amazona agilis*** | | | **RB** | | | |
| Red-crowned Parrot | *Amazona viridigenalis* | | | | | | I |
| Orange-winged Parrot | *Amazona amazonica* | | | | | | I |
| Yellow-crowned Parrot | *Amazona ochrocephala* | I | | | | | I |
| Black-billed Cuckoo | *Coccyzus erythropthalmus* | V | V | V | | V | V |
| Yellow-billed Cuckoo | *Coccyzus americanus* | PM | sb | sb, pm | sb, pm | sb, pm | sb, pm |
| Mangrove Cuckoo | *Coccyzus minor* | RB | rb | RB | RB | RB | RB |
| Great Lizard Cuckoo | *Saurothera (Coccyzus) merlini* | | RB* | | | | |
| **Puerto Rican Lizard Cuckoo** | ***Saurothera (Coccyzus) vieilloti*** | | | | | | **RB** |
| **Hispaniolan Lizard Cuckoo** | ***Saurothera (Coccyzus) longirostris*** | | | | **RB** | **RB** | |
| **Jamaican Lizard Cuckoo** | ***Saurothera (Coccyzus) vetula*** | | | **RB** | | | |
| **Chestnut-bellied Cuckoo** | ***Hyetornis pluvialis*** | | | **RB** | | | |
| **Bay-breasted Cuckoo** | ***Hyetornis rufigularis*** | | | | **rb?** | **RB** | |
| Smooth-billed Ani | *Crotophaga ani* | RB | RB | RB | RB | RB | RB |
| (American) Barn Owl | *Tyto (alba) furcata* | rb | RB | RB | RB | RB | RB |
| **Ashy-faced Owl** | ***Tyto glaucops*** | | | | **RB** | **RB** | |
| **Puerto Rican Screech Owl** | ***Megascops nudipes*** | | | | | | **RB** |
| **Bare-legged Owl** | ***Margarobyas lawrencii*** | | **RB** | | | | |
| **Cuban Pygmy Owl** | ***Glaucidium siju*** | | **RB** | | | | |
| Burrowing Owl | *Speotyto cunicularia* | | rb* | | RB* | RB* | |
| Stygian Owl | *Asio stygius* | | RB* | | rb?* | rb* | |
| Short-eared Owl | *Asio (flammeus) dominguensis* | rb, wv | RB | | rb* | rb* | rb* |
| Long-eared Owl | *Asio otus* | | V | | | | |
| **Jamaican Owl** | ***Pseudoscops grammicus*** | | | **RB** | | | |
| Northern Potoo | *Nyctibius jamaicensis* | | rb (*?) | RB* | rb* | rb* | V |
| Common Nighthawk | *Chordeiles minor* | pm | pm | pm | | pm | ? |
| Antillean Nighthawk | *Chordeiles gundlachii* | SB | SB, PM | SB | SB | SB | SB |
| **Jamaican Poorwill** | ***Siphonorhis americanus*** | | | **E?** | | | |
| **Least Poorwill** | ***Siphonorhis brewsteri*** | | | | **RB** | **RB** | |
| Chuck-will's-widow | *Caprimulgus carolinensis* | pm | wv | wv | WV | WV | wv |
| **Cuban Nightjar** | ***Caprimulgus cubanensis*** | | **RB** | | | | |
| **Hispaniolan Nightjar** | ***Caprimulgus ekmani*** | | | | **rb** | **RB** | |
| Whip-poor-will | *Caprimulgus vociferus* | | V | V | | | |
| **Puerto Rican Nightjar** | ***Caprimulgus noctitherus*** | | | | | | **RB** |
| White-tailed Nightjar | *Caprimulgus cayennensis* | | | | | | V |
| Black Swift | *Cypseloides niger* | V | rb | RB | RB | RB | rb |
| White-collared Swift | *Streptoprocne zonaris* | | rb | RB | RB | RB | V |
| Chimney Swift | *Chaetura pelagica* | PM | pm | V | pm | pm | |
| Short-tailed Swift | *Chaetura brachyura* | | | | | | H |
| Grey-rumped Swift | *Chaetura cinereiventris* | | | | | H | |
| Alpine Swift | *Tachymarptis melba* | | | | | | V |
| Antillean Palm Swift | *Tachornis phoenicobia* | | RB* | RB | RB | RB | V |

| English name | Scientific name | Cayman Islands | Cuba | Jamaica | Haiti | Dominican Republic | Puerto Rico |
|---|---|---|---|---|---|---|---|
| **Jamaican Mango** | ***Anthracothorax mango*** | | | **RB** | | | |
| Green-breasted Mango | *Anthracothorax prevostii* | | | | | | V |
| Antillean Mango | *Anthracothorax dominicus* | | | | RB* | RB* | RB |
| **Green Mango** | ***Anthracothorax viridis*** | | | | | | **RB** |
| Purple-throated Carib | *Eulampis jugularis* | | | | | | V |
| Green-throated Carib | *Eulampis holosericeus* | | | | | | RB |
| Antillean Crested Hummingbird | *Orthorhyncus cristatus* | | | | | | RB |
| Cuban Emerald | *Chlorostilbon ricordii* | | RB* | | | | |
| **Hispaniolan Emerald** | ***Chlorostilbon swainsonii*** | | | | **RB** | **RB** | |
| **Puerto Rican Emerald** | ***Chlorostilbon maugaeus*** | | | | | | **RB** |
| **Black-billed Streamertail** | ***Trochilus scitulus*** | | | **RB** | | | |
| **Red-billed Streamertail** | ***Trochilus polytmus*** | | | **RB** | | | |
| Bahama Woodstar | *Calliphlox evelynae* | | H | | | | |
| Ruby-throated Hummingbird | *Archilochus colubris* | | pm | V | V | V | V |
| Vervain Hummingbird | *Mellisuga minima* | | | RB* | RB* | RB* | V |
| **Bee Hummingbird** | ***Mellisuga helenae*** | | **RB** | | | | |
| **Cuban Trogon** | ***Priotelus temnurus*** | | **RB** | | | | |
| **Hispaniolan Trogon** | ***Priotelus roseigaster*** | | | | **RB** | **RB** | |
| **Cuban Tody** | ***Todus multicolor*** | | **RB** | | | | |
| **Broad-billed Tody** | ***Todus subulatus*** | | | | **RB** | **RB** | |
| **Narrow-billed Tody** | ***Todus angustirostris*** | | | | **RB** | **RB** | |
| **Jamaican Tody** | ***Todus todus*** | | | **RB** | | | |
| **Puerto Rican Tody** | ***Todus mexicanus*** | | | | | | **RB** |
| Common Kingfisher | *Alcedo atthis* | | V | | | | |
| Ringed Kingfisher | *Megaceryle torquata* | | | | | | V |
| Belted Kingfisher | *Megaceryle alcyon* | WV | WV, PM | WV, PM | WV, PM | WV, PM | WV |
| **Antillean Piculet** | ***Nesoctites micromegas*** | | | | **RB** | **RB** | |
| **Puerto Rican Woodpecker** | ***Melanerpes portoricensis*** | | | | | | **RB** |
| **Hispaniolan Woodpecker** | ***Melanerpes striatus*** | | | | **RB** | **RB** | |
| **Jamaican Woodpecker** | ***Melanerpes radiolatus*** | | | **RB** | | | |
| West Indian Woodpecker | *Melanerpes superciliaris* | RB* | RB* | | | | |
| Yellow-bellied Sapsucker | *Sphyrapicus varius* | wv | WV, PM | wv | wv | wv | wv |
| **Cuban Green Woodpecker** | ***Xiphidiopicus percussus*** | | **RB** | **H** | | **H** | |
| Hairy Woodpecker | *Picoides villosus* | | | | | | V |
| Northern Flicker | *Colaptes auratus* | RB* | RB* | | | | |
| **Fernandina's Flicker** | ***Colaptes fernandinae*** | | **RB** | | | | |
| **Ivory-billed Woodpecker** | ***Campephilus principalis*** | | **E?*** | | | | |
| **Jamaican Elaenia** | ***Myiopagis cotta*** | | | **RB** | | | |
| Caribbean Elaenia | *Elaenia martinica* | RB* | | | | | RB |
| Greater Antillean Elaenia | *Elaenia fallax* | | | RB* | RB* | RB* | |
| Western Wood Pewee | *Contopus sordidulus* | | pm | V | | | |
| Eastern Wood Pewee | *Contopus virens* | PM | PM | V | | V | |
| Crescent-eyed Pewee | *Contopus caribaeus* | | RB* | | | | |
| **Hispaniolan Pewee** | ***Contopus hispaniolensis*** | | | | **RB** | **RB** | **V** |
| **Jamaican Pewee** | ***Contopus pallidus*** | | | **RB** | | | |
| **Puerto Rican Pewee** | ***Contopus blancoi*** | | | | | | **RB** |
| Yellow-bellied Flycatcher | *Empidonax flaviventris* | | pm | V | | | |
| Acadian Flycatcher | *Empidonax virescens* | V | pm | | | | |
| Alder Flycatcher | *Empidonax alnorum* | | V | | | | |
| Willow Flycatcher | *Empidonax traillii* | | pm | V | | | |
| Least Flycatcher | *Empidonax minimus* | V | | | | | |
| Eastern Phoebe | *Sayornis phoebe* | V | V | | | | |
| **Sad Flycatcher** | ***Myiarchus barbirostris*** | | | **RB** | | | |
| Great Crested Flycatcher | *Myiarchus crinitus* | | pm | | | V | V |
| **Rufous-tailed Flycatcher** | ***Myiarchus validus*** | | | **RB** | | | |
| La Sagra's Flycatcher | *Myiarchus sagrae* | RB | RB | | | | |
| Stolid Flycatcher | *Myiarchus stolidus* | | | RB* | RB* | RB* | |
| Puerto Rican Flycatcher | *Myiarchus antillarum* | | | | | | RB |

| English name | Scientific name | Cayman Islands | Cuba | Jamaica | Haiti | Dominican Republic | Puerto Rico |
|---|---|---|---|---|---|---|---|
| Tropical Kingbird | *Tyrannus melancholicus* | V | V | | | | |
| Western Kingbird | *Tyrannus verticalis* | | H | | | | |
| Eastern Kingbird | *Tyrannus tyrannus* | PM | pm | V | | | V |
| Grey Kingbird | *Tyrannus dominicensis* | SB | SB, PM | SB | RB | RB | RB |
| Loggerhead Kingbird | *Tyrannus caudifasciatus* | RB* | RB* | RB* | RB* | RB* | RB |
| **Giant Kingbird** | ***Tyrannus cubensis*** | | **RB** | | | | |
| Scissor-tailed Flycatcher | *Tyrannus forficatus* | V | V | | | V | V |
| Fork-tailed Flycatcher | *Tyrannus savana* | V | V | V | | V | |
| **Jamaican Becard** | ***Pachyramphus niger*** | | | **RB** | | | |
| Purple Martin | *Progne subis* | PM | PM | V | | V | V |
| **Cuban Martin** | ***Progne cryptoleuca*** | | **SB** | | | | |
| Caribbean Martin | *Progne dominicensis* | pm | | SB | SB | SB | SB |
| Tree Swallow | *Tachycineta bicolor* | PM | WV, PM | wv, pm | WV, PM? | WV, PM | wv |
| **Golden Swallow** | ***Tachycineta euchrysea*** | | | **E?*** | **rb*** | **rb*** | |
| Bahama Swallow | *Tachycineta cyaneoviridis* | | V | | | | |
| Northern Rough-winged Swallow | *Stelgidopteryx serripennis* | PM | PM | wv, pm | wv, pm | wv, pm | |
| Bank Swallow | *Riparia riparia* | PM | pm | pm | pm | pm | pm |
| Cliff Swallow | *Petrochelidon pyrrhonota* | pm | pm | V | V | V | |
| Cave Swallow | *Petrochelidon fulva* | pm | SB* | RB* | RB* | RB* | RB |
| Barn Swallow | *Hirundo rustica* | WV, PM | PM | WV, PM | WV, PM | WV, PM | wv, PM |
| **Cuban Palm Crow** | ***Corvus minutus*** | | **rb** | | | | |
| **Hispaniolan Palm Crow** | ***Corvus palmarum*** | | | | **RB** | **RB** | |
| Cuban Crow | *Corvus nasicus* | | RB | | | | |
| **White-necked Crow** | ***Corvus leucognaphalus*** | | | | **RB** | **RB** | **E** |
| **Jamaican Crow** | ***Corvus jamaicensis*** | | | **RB** | | | |
| House Crow | *Corvus splendens* | | H | | | | |
| **Zapata Wren** | ***Ferminia cerverai*** | | **rb** | | | | |
| House Wren | *Troglodytes aedon* | | V | | | | |
| Marsh Wren | *Cistothorus palustris* | | V | | | | |
| Ruby-crowned Kinglet | *Regulus calendula* | | V | V | | V | |
| Blue-grey Gnatcatcher | *Polioptila caerulea* | wv, pm | WV | | | V | |
| **Cuban Gnatcatcher** | ***Polioptila lembeyei*** | | **RB** | | | | |
| Northern Wheatear | *Oenanthe oenanthe* | | V | | | | V |
| Eastern Bluebird | *Sialia sialis* | | V | | | | |
| **Cuban Solitaire** | ***Myadestes elisabeth*** | | **RB** | | | | |
| Rufous-throated Solitaire | *Myadestes genibarbis* | | | RB* | RB* | RB* | |
| Veery | *Catharus fuscescens* | pm | pm | pm | | V | V |
| Grey-cheeked Thrush | *Catharus minimus* | pm | pm | ? | | | pm |
| Bicknell's Thrush | *Catharus bicknelli* | | wv | wv | WV | WV | wv |
| Swainson's Thrush | *Catharus ustulatus* | PM | pm | wv | | V | V |
| Hermit Thrush | *Catharus guttatus* | | V | | | | |
| Wood Thrush | *Hylocichla mustelina* | pm | V | V | | V | V |
| **White-eyed Thrush** | ***Turdus jamaicenis*** | | | **RB** | | | |
| American Robin | *Turdus migratorius* | | pm | V | | V | |
| **La Selle Thrush** | ***Turdus swalesi*** | | | | **RB*** | **RB*** | |
| **White-chinned Thrush** | ***Turdus aurantius*** | | | **RB** | | | |
| **Grand Cayman Thrush** | ***Turdus ravidus*** | **E** | | | | | |
| Red-legged Thrush | *Turdus plumbeus* | RB* | RB | | RB* | RB* | RB* |
| Grey Catbird | *Dumetella carolinensis* | WV | WV, PM | wv | wv | wv | V |
| Northern Mockingbird | *Mimus polyglottos* | RB | RB | RB | RB | RB | RB |
| Bahama Mockingbird | *Mimus gundlachii* | | rb | rb* | | | |
| Brown Thrasher | *Toxostoma rufum* | | V | | | | |
| Pearly-eyed Thrasher | *Margarops fuscatus* | | | V | | rb | RB |
| American (Buff-bellied) Pipit | *Anthus rubescens* | | V | V | V | | |
| Cedar Waxwing | *Bombycilla cedrorum* | WV, PM | WV | wv | wv | wv | V |
| **Palmchat** | ***Dulus dominicus*** | | | | **RB** | **RB** | |
| European Starling | *Sturnus vulgaris* | V | V | rb | | | V |

| English name | Scientific name | Cayman Islands | Cuba | Jamaica | Haiti | Dominican Republic | Puerto Rico |
|---|---|---|---|---|---|---|---|
| Hill Myna | *Gracula religiosa* | | | | | | I |
| White-eyed Vireo | *Vireo griseus* | wv, pm | WV | wv | V | V | V |
| Thick-billed Vireo | *Vireo crassirostris* | RB* | rb* | | rb* | | |
| **Jamaican Vireo** | ***Vireo modestus*** | | | **RB** | | | |
| **Cuban Vireo** | ***Vireo gundlachii*** | | **RB** | | | | |
| **Puerto Rican Vireo** | ***Vireo latimeri*** | | | | | | **RB** |
| **Flat-billed Vireo** | ***Vireo nanus*** | | | | **RB** | **RB** | |
| **Blue Mountain Vireo** | ***Vireo osburni*** | | | **RB** | | | |
| Blue-headed Vireo | *Vireo solitarius* | | wv | V | | | |
| Yellow-throated Vireo | *Vireo flavifrons* | WV | wv | V | V | V | V |
| Warbling Vireo | *Vireo gilvus* | | wv, pm | V | | V | |
| Philadelphia Vireo | *Vireo philadelphicus* | pm | wv, pm | V | | | V |
| Red-eyed Vireo | *Vireo olivaceus* | PM | PM | pm | | V | V |
| Black-whiskered Vireo | *Vireo altiloquus* | SB, PM | SB | SB | SB, wv | SB, wv | SB |
| Yucatan Vireo | *Vireo magister* | RB* | | | | | |
| Bachman's Warbler | *Vermivora bachmanii* | | E? | | | | |
| Blue-winged Warbler | *Vermivora pinus* | wv, pm | wv, pm | wv | wv | wv | wv |
| Golden-winged Warbler | *Vermivora chrysoptera* | pm | pm | V | | wv | wv |
| Tennessee Warbler | *Vermivora (Leiothlypis) peregrina* | wv, pm | pm | wv | | wv | |
| Orange-crowned Warbler | *Vermivora (Leiothlypis) celata* | H | V | V | | | |
| Nashville Warbler | *Vermivora (Leiothlypis) ruficapilla* | pm | V | V | V | V | V |
| Virginia's Warbler | *Vermivora (Leiothlypis) virginiae* | | H | | | | |
| Northern Parula | *Parula americana* | WV | WV, PM | WV | WV | WV | WV |
| Yellow Warbler | *Dendroica petechia* | RB, WV, PM | RB, WV | RB, wv | RB*, wv | RB*, wv | RB, wv |
| Chestnut-sided Warbler | *Dendroica pensylvanica* | PM | wv, pm | pm | H | V | V |
| Magnolia Warbler | *Dendroica magnolia* | wv, pm | WV, PM | wv | WV | WV | wv |
| Cape May Warbler | *Dendroica tigrina* | WV | WV, PM | WV | WV | WV | WV |
| Black-throated Blue Warbler | *Dendroica caerulescens* | WV | WV, PM | WV | WV | WV | WV |
| Yellow-rumped Warbler | *Dendroica coronata* | wv, pm | WV, PM | WV, PM | WV, PM | WV, PM | wv |
| Townsend's Warbler | *Dendroica townsendi* | | | | | | H |
| Black-throated Grey Warbler | *Dendroica nigrescens* | | V | | | | |
| Black-throated Green Warbler | *Dendroica virens* | wv, pm | wv, pm | wv | wv | wv | wv |
| Blackburnian Warbler | *Dendroica fusca* | pm | pm | wv | | V | V |
| Yellow-throated Warbler | *Dendroica dominica* | WV | WV, PM | wv | WV | WV | wv |
| **Adelaide's Warbler** | ***Dendroica adelaidae*** | | | | | | **RB** |
| Olive-capped Warbler | *Dendroica pityophila* | | RB* | | | | |
| Pine Warbler | *Dendroica pinus* | pm | V | V | RB* | RB* | V |
| Kirtland's Warbler | *Dendroica kirtlandii* | | H | | | H | |
| Prairie Warbler | *Dendroica discolor* | WV | WV, PM | WV | WV | WV | WV |
| Vitelline Warbler | *Dendroica vitellina* | RB* | | | | | |
| Palm Warbler | *Dendroica palmarum* | WV, PM | WV, PM | WV, PM | WV | WV | WV |
| Bay-breasted Warbler | *Dendroica castanea* | Pm | pm | wv | V | pm | V |
| Blackpoll Warbler | *Dendroica striata* | Pm | PM | pm | PM | PM | pm |
| Cerulean Warbler | *Dendroica cerulea* | wv, pm | V | V | | | V |
| **Arrowhead Warbler** | ***Dendroica pharetra*** | | | **RB** | | | |
| **Elfin Woods Warbler** | ***Dendroica angelae*** | | | | | | **RB** |
| Black-and-white Warbler | *Mniotilta varia* | WV, PM | WV, PM | WV | WV | WV | WV |
| American Redstart | *Setophaga ruticilla* | WV, PM | WV, PM | WV, PM | WV, PM | WV, PM | WV |
| Prothonotary Warbler | *Protonotaria citrea* | wv, pm | pm | wv, pm | | wv, pm | pm |
| Worm-eating Warbler | *Helmitheros vermivorum* | WV | WV, PM | wv | wv | wv | wv |
| Swainson's Warbler | *Limnothlypis swainsonii* | pm | wv, pm | wv | V | wv | V |
| Ovenbird | *Seiurus aurocapilla* | WV, PM | WV, PM | WV | WV | WV | WV |
| Northern Waterthrush | *Seiurus (Parkesia) noveboracensis* | WV, PM | WV, PM | WV, PM | WV, PM | WV, PM | WV |
| Louisiana Waterthrush | *Seiurus (Parkesia) motacilla* | pm | WV, PM | wv, pm | wv, pm | wv, pm | WV |

| English name | Scientific name | Cayman Islands | Cuba | Jamaica | Haiti | Dominican Republic | Puerto Rico |
|---|---|---|---|---|---|---|---|
| Kentucky Warbler | *Oporornis formosus* | wv, pm | wv, pm | V | | V | V |
| Connecticut Warbler | *Oporornis agilis* | H | | V | | pm | V |
| Mourning Warbler | *Oporornis philadelphia* | H | V | V | | V | V |
| Common Yellowthroat | *Geothlypis trichas* | WV, PM? | WV, PM | WV, PM | WV, PM | WV, PM | WV |
| **Yellow-headed Warbler** | ***Teretistris fernandinae*** | | **RB** | | | | |
| **Oriente Warbler** | ***Teretistris fornsi*** | | **RB** | | | | |
| Hooded Warbler | *Wilsonia citrina* | wv, pm | wv, pm | wv, pm | | wv, pm | V |
| Wilson's Warbler | *Wilsonia pusilla* | pm | wv, pm | V | | V | V |
| Canada Warbler | *Wilsonia canadensis* | pm | pm | V | | V | V |
| Yellow-breasted Chat | *Icteria virens* | V | pm | V | | V | |
| Bananaquit | *Coereba flaveola* | RB* | V | RB* | RB* | RB* | RB |
| Red-legged Honeycreeper | *Cyanerpes cyaneus* | | RB (I?) | | | | |
| **Jamaican Euphonia** | ***Euphonia jamaica*** | | | **RB** | | | |
| Antillean Euphonia | *Euphonia musica* | | | | RB* | RB* | RB* |
| Western Spindalis | *Spindalis zena* | RB* | RB* | | | | |
| **Jamaican Spindalis** | ***Spindalis nigricephalus*** | | | **RB** | | | |
| **Hispaniolan Spindalis** | ***Spindalis dominicensis*** | | | | **RB** | **RB** | |
| **Puerto Rican Spindalis** | ***Spindalis portoricensis*** | | | | | | **RB** |
| **Green-tailed Ground Tanager** | ***Microligea palustris*** | | | | **RB** | **RB** | |
| **Hispaniolan Highland Tanager** | ***Xenoligea montana*** | | | | **RB** | **RB** | |
| Summer Tanager | *Piranga rubra* | wv, pm | wv, pm | wv, pm | | V | V |
| Scarlet Tanager | *Piranga olivacea* | wv, pm | pm | wv, pm | | V | pm |
| Western Tanager | *Piranga ludoviciana* | | H | | | V | |
| **Black-crowned Palm Tanager** | ***Phaenicophilus palmarum*** | | | | **RB** | **RB** | |
| **Grey-crowned Palm Tanager** | ***Phaenicophilus poliocephalus*** | | | | **RB** | **H** | |
| **Eastern Chat-Tanager** | ***Calyptophilus frugivorus*** | | | | **RB** | **RB** | |
| **Western Chat-Tanager** | ***Calyptophilus tertius*** | | | | **RB** | **RB** | |
| **Puerto Rican Tanager** | ***Nesospingus speculiferus*** | | | | | | **RB** |
| Swallow Tanager | *Tersina viridis* | I | | | | | |
| Rose-breasted Grosbeak | *Pheucticus ludovicianus* | WV, PM | wv, pm | wv, pm | wv, pm | wv, pm | wv |
| Black-headed Grosbeak | *Pheucticus melanocephalus* | | V | | | | |
| Blue Grosbeak | *Passerina caerulea* | pm | wv, pm | V | | wv | V |
| Indigo Bunting | *Passerina cyanea* | WV, PM | WV, PM | wv | | wv | wv |
| Lazuli Bunting | *Passerina amoena* | | V | | | | |
| Painted Bunting | *Passerina ciris* | wv, pm | WV, PM | V | | V | |
| Dickcissel | *Spiza americana* | | pm | V | | | V |
| Cuban Bullfinch | *Melopyrrha nigra* | RB* | RB* | | | | |
| **Cuban Grassquit** | ***Tiaris canorus*** | | **RB** | | | | |
| Yellow-faced Grassquit | *Tiaris olivaceus* | RB | RB | RB | RB | RB | RB |
| Black-faced Grassquit | *Tiaris bicolor* | | rb | RB | RB | RB | RB |
| **Yellow-shouldered Grassquit** | ***Loxipasser anoxanthus*** | | | **RB** | | | |
| **Puerto Rican Bullfinch** | ***Loxigilla portoricensis*** | | | | | | **RB** |
| Greater Antillean Bullfinch | *Loxigilla violacea* | | | RB* | RB* | RB* | |
| **Orangequit** | ***Euneornis campestris*** | | | **RB** | | | |
| Saffron Finch | *Sicalis flaveola* | | V | RB | | I? | I |
| Red-crested Cardinal | *Paoraria coronata* | | | | | | I |
| Green-tailed Towhee | *Pipilo chlorurus* | | V | | | | |
| **Zapata (Cuban) Sparrow** | ***Torreornis inexpectata*** | | **RB** | | | | |
| Chipping Sparrow | *Spizella passerina* | | V | | | | |
| Clay-coloured Sparrow | *Spizella pallida* | | wv, pm | | | | |
| Lark Sparrow | *Chondestes grammacus* | | V | V | | | |
| Savannah Sparrow | *Passerculus sandwichensis* | pm | wv, pm | | V | | |
| Grasshopper Sparrow | *Ammodramus savannarum* | pm | wv, pm | rb* | rb* | rb* | RB |
| Song Sparrow | *Melospiza melodia* | | | | | V | |
| Lincoln's Sparrow | *Melospiza lincolnii* | | wv, pm | wv | V | | V |
| Rufous-collared Sparrow | *Zonotrichia capensis* | | | | | RB* | |
| White-throated Sparrow | *Zonotrichia albicollis* | | | | | | V |

| English name | Scientific name | Cayman Islands | Cuba | Jamaica | Haiti | Dominican Republic | Puerto Rico |
|---|---|---|---|---|---|---|---|
| White-crowned Sparrow | *Zonotrichia leucophrys* | V | wv, pm | V | | | |
| Dark-eyed Junco | *Junco hyemalis* | | H | V | | | V |
| Bobolink | *Dolichonyx oryzivorus* | PM | pm | | pm | pm | pm |
| **Red-shouldered Blackbird** | ***Agelaius assimilis*** | | **rb** | | | | |
| Tawny-shouldered Blackbird | *Agelaius humeralis* | | RB* | | RB* | H | |
| **Yellow-shouldered Blackbird** | ***Agelaius xanthomus*** | | | | | | **RB** |
| **Jamaican Blackbird** | ***Nesopsar nigerrimus*** | | | **rb** | | | |
| Eastern Meadowlark | *Sturnella magna* | V | RB* | | | | |
| Yellow-headed Blackbird | *Xanthocephalus xanthocephalus* | V | V | | | | V |
| **Cuban Blackbird** | ***Dives atroviolaceus*** | | **RB** | | | | |
| Rusty Blackbird | *Euphagus carolinus* | | H | | | | |
| Greater Antillean Grackle | *Quiscalus niger* | RB* | RB* | RB* | RB* | RB* | RB* |
| Shiny Cowbird | *Molothrus bonariensis* | ? | RB | rb | RB | RB | RB |
| Brown-headed Cowbird | *Molothrus ater* | | V | | | | |
| **Cuban Oriole** | ***Icterus melanopsis*** | | **RB** | | | | |
| **Hispaniolan Oriole** | ***Icterus dominicensis*** | | | | **RB** | **RB** | |
| **Puerto Rican Oriole** | ***Icterus portoricensis*** | | | | | | **RB** |
| Orchard Oriole | *Icterus spurius* | | wv, pm | V | | V | |
| Hooded Oriole | *Icterus cucullatus* | | V | | | | |
| Venezuelan Troupial | *Icterus icterus* | | | | | | I |
| Jamaican Oriole | *Icterus leucopteryx* | E* | | RB* | | H | |
| Audubon's Oriole | *Icterus graduacuda* | | | | | | I? |
| Baltimore Oriole | *Icterus galbula* | wv, pm | wv, pm | wv, pm | wv, pm | wv, pm | wv |
| **Hispaniolan Crossbill** | ***Loxia megaplaga*** | | | **V** | **RB** | **RB** | |
| Common Redpoll | *Carduelis flammea* | | | V? | | | |
| Red Siskin | *Carduelis cucullata* | | | | | | I |
| **Antillean Siskin** | ***Carduelis dominicensis*** | | | | **RB** | **RB** | |
| American Goldfinch | *Carduelis tristis* | | V | | | | |
| Lesser Goldfinch | *Carduelis psaltria* | | I (E) | | | | |
| Yellow-fronted Canary | *Serinus mozambicus* | | | | | | I |
| House Sparrow | *Passer domesticus* | | I | I (V) | I | I | I |
| Village Weaver | *Ploceus cucullatus* | | | | I | I | |
| Orange Bishop | *Euplectes franciscanus* | | | I (V) | | | I |
| Yellow-crowned Bishop | *Euplectes afer* | | | I | | | I |
| Orange-cheeked Waxbill | *Estrilda melpoda* | | | | | | I |
| Black-rumped Waxbill | *Estrilda troglodytes* | | | | | | I |
| Red Avadavat | *Amandava amandava* | | | | | I | I |
| Indian Silverbill | *Lonchura malabarica* | | | | | | I |
| Bronze Mannikin | *Lonchura cucullata* | | | | | | I |
| **Nutmeg Mannikin** | *Lonchura punctulata* | | I | I | I | I | I |
| Chestnut Mannikin | *Lonchura malacca* | | I | I | I | I | I |
| Java Sparrow | *Padda oryzivora* | | | | | | I |
| Pin-tailed Whydah | *Vidua macroura* | | | | | | I |

# MAMMALS OF THE GREATER ANTILLES

**Checklist** Nomenclature follows *Mammals of the world: a checklist* (Duff and Lawson)

**Abbreviations used:**

| **Status** | | **Range** | |
|---|---|---|---|
| CR | Critically Endangered | C | Cuba |
| EN | Endangered | D | Dominican Republic |
| VU | Vulnerable | H | Haiti |
| NT | Near Threatened | J | Jamaica |
| LC | Least Concern | P | Puerto Rico |
| DD | Data Deficient | | |

**Order: Rodentia (Rodents)**

| | | | |
|---|---|---|---|
| Coues' Rice Rat | *Oryzomys couesi* | LC | J |
| Desmarest's Hutia | *Capromys pilorides* | LC | C |
| Cabrera's Hutia | *Mesocapromys angelcabrerai* | CR | C |
| Eared Hutia | *Mesocapromys auritus* | CR | C |
| Dwarf Hutia | *Mesocapromys nanus* | CR | C |
| San Felipe Hutia | *Mesocapromys sanfelipensis* | CR | C |
| Garrido's Hutia | *Mysateles garridoi* | CR | C |
| Gundlach's Hutia | *Mysateles gundlachi* | VU | C |
| Black-tailed Hutia | *Mysateles melanurus* | NT | C |
| Southern Hutia | *Mysateles meridionalis* | NT | C |
| Prehensile-tailed Hutia | *Mysateles prehensilis* | LC | C |
| Puerto Rican Hutia | *Isobolodon portoricensis* | CR | D, H, P |
| Brown's Hutia | *Geocapromys brownii* | VU | J |
| Hispaniolan Hutia | *Plagiodontia aedium* | VU | D, H |

**Order: Carnivora (Carnivores)**

| | | | |
|---|---|---|---|
| Hooded Seal | *Cystophora cristata* | LC | P |

**Order: Soricomorpha (Shrews, Moles and Solenodons)**

| | | | |
|---|---|---|---|
| Cuban Solenodon | *Solenodon cubanus* | EN | C |
| Hispaniolan Solenodon | *Solenodon paradoxus* | EN | D, H |

**Order: Chiroptera (Bats)**

| | | | |
|---|---|---|---|
| Greater Bulldog Bat | *Noctilio leorinus* | LC | C, D, H, J, P |
| Macleay's Moustached Bat | *Pteronotus macleayii* | VU | C, J |
| Sooty Moustached Bat | *Pteronotus quadridens* | NT | C, D, H, J, P |
| Parnell's Moustached Bat | *Pteronotus parnellii* | LC | C, D, H, J, P |
| Antillean Ghost-faced Bat | *Mormoops blainvillii* | NT | C, D, H, J, P |
| Spectral Bat | *Vampyrum spectrum* | NT | J |
| Antillean Fruit-eating Bat | *Brachyphylla cavernarum* | LC | P |

**Order: Chiroptera (Bats)** ***(continued)***

| | | | |
|---|---|---|---|
| Waterhouse's Leaf-nosed Bat | *Macrotus waterhousii* | LC | C, D, H, J, P |
| Cuban Fruit-eating Bat | *Brachyphylla nana* | NT | C, D, H, J |
| Buffy Flower Bat | *Erophylla sezekorni* | LC | C, J |
| Cuban Flower Bat | *Phyllonycteris poeyi* | NT | C, H |
| Jamaican Flower Bat | *Phyllonycteris aphylla* | EN | J |
| Common Long-tongued Bat | *Glossophaga sorcina* | LC | J |
| Leach's Single Leaf Bat | *Monophyllus redmani* | LC | C, D, H, J, P |
| Seba's Short-tailed Bat | *Carollia perspicillata* | LC | J |
| Little Yellow-shouldered Bat | *Sturnira lilium* | LC | J |
| Jamaican Fruit Bat | *Artibeus jamaicensis* | LC | C, D, H, J, P |
| Cuban Fig-eating Bat | *Phyllops falcatus* | NT | C, D, H |
| Jamaican Fig-eating Bat | *Ariteus flavescens* | VU | J |
| Cuban Funnel-eared Bat | *Natalus micropus* | LC | C, D, H, J |
| Mexican Funnel-eared Bat | *Natalus stramineus* | LC | C, D, H, J |
| Gervais's Funnel-eared Bat | *Natalus lepidus* | NT | C |
| Little Goblin Bat | *Mormopterus minutus* | VU | C |
| Pallas' Free-tailed Bat | *Molossus molossus* | LC | C, D, H, J, P |
| Mexican Free-tailed Bat | *Tadarida brasiliensis* | NT | C, D, H, J, P |
| Broad-eared Bat | *Nyctinomops laticaudatus* | LC | C |
| Big Free-tailed Bat | *Nyctinomops macrotis* | LC | C, D, H, J |
| Red Fruit Bat | *Stenoderma rufum* | VU | P |
| Wagner's Bonneted Bat | *Eumpos glaucinus* | LC | C, J |
| Greater Bonneted Bat | *Eumpos perotis* | LC | C |
| Pallid Bat | *Antrozus pallidus* | LC | C |
| Big Brown Bat | *Eptesicus fuscus* | LC | C |
| Northern Yellow Bat | *Lasiurus intermedius* | LC | C |
| Eastern Red Bat | *Lasiurus borealis* | LC | C, D, H, J, P |
| Vesper Bat | *Nycticeius humeralis* | LC | C |

**Order: Cetacea**

| | | | |
|---|---|---|---|
| Sei Whale | *Balaenoptera borealis* | EN | C |
| Humpback Whale | *Megaptera novaeangliae* | VU | D |
| Pygmy Sperm Whale | *Kogia breviceps* | LC | C, H, P |
| Dwarf Pygmy Whale | *Kogia sima* | LC | C, D |
| Gervais' Beaked Whale | *Mesoplodon europaeus* | DD | C, J |
| True's Beaked Whale | *Mesoplodon mirus* | DD | C |
| Cuvier's Beaked Whale | *Ziphius cavirostris* | DD | C |
| Killer Whale | *Orcinus orca* | NT | C, J |
| False Killer Whale | *Pseudorca crassidens* | LC | C |
| Common Bottle-nosed Dolphin | *Tursiops truncatus* | DD | P |
| Rough-toothed Dolphin | *Steno bredanensis* | DD | C, D, J, P |
| Fraser's Dolphin | *Lagenodelphis hosei* | DD | C, D, H, J, P |
| Clymene Dolphin | *Stenella clymene* | DD | C, D, H, J, P |
| Striped Dolphin | *Stenella coeruleoalba* | NT | C, J |
| Atlantic Spotted Dolphin | *Stenella frontalis* | DD | H, P |
| Spinner Dolphin | *Stenella longirostris* | NT | C, H, J |
| Short-finned Pilot Whale | *Globicephala macrorynhus* | NT | H, P |
| Risso's Dolphin | *Grampus griseus* | DD | C, D, H, J, P |

**Order: Cetacea *(continued)***

| | | | |
|---|---|---|---|
| Pygmy Killer Whale | *Feresa attenuata* | DD | C, D, J |

**Order: Sirenia (Manatees and Dugongs)**

| | | | |
|---|---|---|---|
| West Indian Manatee | *Trichechus manatus* | VU | C, D, H, J, P |

## References


Baillie, J. E. M. (ed.) 2004. *IUCN Red List of threatened species: a global species assessment*. International Union for the Conservation of Nature and Natural Resources, Gland.

Duff, A. and A. Lawson. 2004. *Mammals of the world: a checklist*. Christopher Helm, London.

Wilson, D. E. and D. E. M. Reeder. (eds.) 2005. *Mammal species of the world, a taxonomic and geographic reference*. 2 volumes. Johns Hopkins University Press, Baltimore.

# USEFUL ADDRESSES, SOCIETIES AND CLUBS

*Tour operators*

There are now more companies than ever offering specialist birding holidays in Cuba and elsewhere in the Greater Antilles. Given current constraints on US travel and links with Cuba, it is no surprise that the majority of those offering tours of that island are based in Europe, although an increasing number of bodies in North America are organising legal trips to the island. The following is a list of those companies that we are aware of, and can recommend, that are currently operating birding tours to Cuba and the Greater Antilles (although not necessarily on an annual basis). We should emphasise that these are holidays for serious birders, whose aim is to see as many as possible of the endemics, not mixed-interest tours, of which one element is birding. Readers of this book, we feel sure, will be seeking only the former type of trip. In alphabetical order they are: Birdfinders (www.birdfinders.co.uk), Birdquest (www.birdquest.co.uk), Birdseekers (www.birdseekers.com), Birdwatching Breaks (www.birdwatching breaks.com), Eagle Eye Tours (www.eagle-eye.com), Island Holidays (www.islandholidays.co.uk), Limosa (www.limosaholidays.co.uk), Ornitholidays (www.ornitholidays.co.uk) and Sunbird (www.sunbirdtours.co.uk). This is not intended to be an exhaustive list, it is intended as a guide to those companies that one or more of us have either worked with or can recommend through other personal knowledge.

*Specialist Cuban travel agencies*

There are now many travel agencies and tour operators offering specialist-booking arrangements for those travelling to and within Cuba. Some of these also offer knowledge of organising guided birdwatching holidays and/or provide ground arrangements for the companies mentioned in the previous paragraph. In general we recommend that if you want a guided birding tour you use one of the ten specialist operators indicated above, or another of which you have personal, positive experience. While in some instances you may find that a regular high-street travel agent can make all of your necessary bookings on your behalf, we consider that there are a certain number of advantages, some of them considerable, to making your arrangements through an operator specialising in holidays to Cuba. In this respect we particularly recommend Havanatour, which has an office in the UK (3 Wyllyotts Place, Potters Bar, Herts EN6 2JD; telephone: 01707 646463; e-mail: sales@havanatour.co.uk), and Cubatur, which has an agent in the Republic of Ireland (telephone: 016 713422). Both have considerable experience in organising birding tours for the major bird tour companies, as well as helping solo travellers create their own itinerary. Advantages to booking through one of these agencies include their access to the cheapest airline fares on routes to Cuba, with the tourist card often included in the price of your ticket, favourable rates on car hire, and good deals on hotels throughout the island. Although we have not used them, Cubancán is among the major Cuban tour agencies and also has a UK arm (telephone: 020 75368173; www.cubanacan.co.uk).

If you wish to make your own arrangements, either in advance or on a daily basis while in Cuba, you may find the following websites worthy of checking. They have addresses of hotels, tour agencies, car hire and other services: www.cubawelcome.com, www.dtcuba.com, www.cubaweb.cu, www.easytravel.cu and www.travelnet.cu.

***Birding organisations***

Those with an interest in Cuban or West Indian birds should consider supporting one or both of the following organisations. The **Society for the Conservation and Study of Caribbean Birds** (SCSCB) publishes the thrice-annual *Journal of Caribbean Ornithology*, which is packed with new information from studies in the region, and also organises regular ornithological conferences in the West Indian region, several of which have been held in the Greater Antilles. The annual membership costs are US$20 for individuals, US$120 for US-based institutions and US$50 for Caribbean-based institutions. Membership dues should be sent to Rosemarie Gnam, Treasurer SCSCB, P.O. Box 863208, Ridgewood, NY 11386, USA. Back issues can be viewed online: http://web2.puc.edu/Faculty/Floyd_Hayes/jco/archives.html. Further information about the Society can be found at: www.nmnh.si.edu/BIRDNET/SCO/index.html.

In addition, the **Neotropical Bird Club** (NBC) publishes the biannual colour magazine *Neotropical Birding* and the annual journal *Cotinga*, which regularly features papers on West Indian birds, as well as news of conservation developments, novel taxonomic studies and interesting new bird records from the region. NBC has developed a particularly strong association with Cuban ornithology, and has funded a number of nationally operated conservation projects in the country through its Conservation Awards scheme. The Club also has birders' trip reports for sale (profits are used to fund Conservation Awards) covering many countries in the Neotropics, including Cuba. Dues are currently UK£21/US$40 for individuals, US$25 for nationals resident in the region, and UK£45/US$90 for organisations, libraries and institutions. Further information about NBC, including how to join can be viewed at www.neotropicalbirdclub.org. Alternatively contact NBC, c/o The Lodge, Sandy, Beds SG19 2DL, or in the USA via The American Birding Association, P.O. Box 6599, Colorado Springs, CO 80934.

Interesting bird records from throughout the Greater Antilles will be welcomed by the **West Indies regional editor** for *North American Birds*, Robert L. Norton, who can be contacted at 8690 NE Waldo Road, Gainesville, FL 32609, USA, e-mail: rnorton@co.alachua.fl.us or corvus0486@aol.com.

# SELECTED BIBLIOGRAPHY 

**Barbour, T.** 1923. *The birds of Cuba*. Memoirs of the Nuttall Ornithological Club No. 4. Cambridge, Massachusetts.

**Barlow, J. C.** 1990. *Songs of the vireos and their allies*. Revised edn. Cassette. ARA Records, Gainesville, Florida.

**BirdLife International.** 2004. *Threatened birds of the world 2004*. CD-ROM. BirdLife International, Cambridge, UK.

**BirdLife International.** 2008. *Important Bird Areas in the Caribbean: key sites for conservation*. BirdLife International, Cambridge, UK.

**Bradley, P. E.** 2000. *The birds of the Cayman Islands*. BOU Checklist 19. British Ornithologists' Union, Tring.

**Dod, A. S.** 1978. *Aves de la Republica Dominicana*. Museo Nacional de Historia Natural, Santo Domingo.

**Downer, A. and R. Sutton.** 1990. *Birds of Jamaica: a photographic field guide*. Cambridge University Press, Cambridge, UK.

**Flieg, G. M. and A. Sander.** 2000. *A photographic guide to the birds of the West Indies*. New Holland, London.

**Garrido, O. and A. Kirkconnell.** 2000. *Field guide to the birds of Cuba*. Cornell University Press, Ithaca, New York.

**González Alonso, H.** 2002. *Aves de Cuba*. Instituto de Ecología y Sistemática, La Habana.

**Hardy, J. W., Reynard, G. B. and Coffey, B. B.** 1989. *Voices of the New World pigeons and doves*. Cassette. ARA Records, Gainesville, Florida.

**Haynes Sutton, A., A. Downer and R. Sutton.** 2009. *A photographic guide to the birds of Jamaica*. Christopher Helm, London.

**Keith, A. R., J. W. Wiley, S. C. Latta and J. A. Ottenwalder.** 2003. *The birds of Hispaniola*. BOU Checklist 21. British Ornithologists' Union & British Ornithologists' Club, Tring.

**Kirkconnell, A., G. M. Kirwan, W. Suárez, J. Wiley, A. Mitchell and O. Garrido.** In prep. *The birds of Cuba: an annotated checklist*. British Ornithologists' Union & British Ornithologists' Club, Peterborough.

**Latta, S. C.** (ed.) 1998. *Recent ornithological research in the Dominican Republic / Investigaciones ornitológicas recientes en la República Dominicana*. Ediciones Tinglar, Santo Domingo.

**Latta, S., C. Rimmer, A. Keith, J. Wiley, H. Raffaele, K. McFarland and E. Fernandez.** 2006. *Birds of the Dominican Republic and Haiti*. Christopher Helm, London.

**Nellis, D. W.** 2000. *Puerto Rico and Virgin Islands wildlife viewing guide.* Falcon Publishing, Helena, Montana.

**Oberle, M. W.** 2000. *Puerto Rico's birds in photographs.* Second edition. Book and CD-ROM. Editorial Humanitas, San Juan.

**Ponce, J. A. A., L. B. Mesa, L. F. Pérez, O. Martínez, T. P. Corbelo and G. Forneris.** 2002. *La Ciénaga de Zapata: historia y naturaleza.* NAG, Turin.

**Raffaele, H. A.** 1989. *A guide to the birds of Puerto Rico and the Virgin Islands.* Princeton University Press, Princeton, New Jersey.

**Raffaele, H., J. Wiley, O. Garrido, A. Keith and J. Raffaele.** 1998 and 2003. *Birds of the West Indies.* Christopher Helm, London.

**Reynard, G. B. and O. H. Garrido.** 2006. *Bird songs in Cuba.* CDs. Library of Natural Sounds, Cornell Laboratory of Ornithology, Ithaca, New York.

**Reynard, G. B. and R. L. Sutton.** 2000. *Bird songs in Jamaica.* CDs. Library of Natural Sounds, Cornell Laboratory of Ornithology, Ithaca, New York.

**Silva Taboada, G., W. Suárez Duque and S. Díaz Franco.** 2008. *Compendio de los mamíferos terrestres autóctonos de Cuba.* Museo Nacional de Historia Natural, La Habana.

**Stattersfield, A. J., M. J. Crosby, A. J. Long and D. C. Wege.** 1998. *Endemic Bird Areas of the world: priorities for biodiversity conservation.* BirdLife International, Cambridge, UK.

**Wetmore, A. and B. H. Swales.** 1931. *The birds of Haiti and the Dominican Republic.* US National Museum Bulletin 155. Smithsonian Institution, Washington DC.

**Wauer, R. H.** 1996. *A birder's West Indies: an island-by-island tour.* University of Texas Press, Austin.

**Wheatley, N. and D. Brewer.** 2001. *Where to watch birds in Central America and the Caribbean.* Christopher Helm, London.

**Whitney, B. M., Parker, T. A., Budney, G. F., Munn, C. A. and Bradbury, J. W.** 2002. *Voices of the New World parrots.* CDs. Macaulay Library of Natural Sounds, Cornell Laboratory of Ornithology, Ithaca, New York.